High-Energy Ecologically Safe HF/DF Lasers

High-Energy Ecologically Safe HF/DF Lasers

Physics of Self-Initiated
Volume Discharge-Based HF/DF Lasers

Victor V. Apollonov
Sergey Yu. Kazantsev

CRC Press
Taylor & Francis Group
Boca Raton London New York

CISP

CRC Press is an imprint of the
Taylor & Francis Group, an **informa** business

Translated from Russian by V.E. Riecansky

CRC Press
Taylor & Francis Group
6000 Broken Sound Parkway NW, Suite 300
Boca Raton, FL 33487-2742

First issued in paperback 2021

© 2020 by CISP
CRC Press is an imprint of Taylor & Francis Group, an Informa business

No claim to original U.S. Government works

ISBN 13: 978-1-03-223770-1 (pbk)
ISBN 13: 978-0-367-47820-9 (hbk)

The book is dedicated to the 100-years jubilee of our colleague prof. A. I. Barchukov.

Contents

Introduction

Relevance of the work. At present, in the spectral region of 2.6-4.1 μm, according to the energy characteristics (laser energy in a pulse, average and pulsed power) and convenience of operation, chemical HF(DF) lasers on the non-chain reaction initiated by a self-sustained volume discharge (SSVD), are significantly superior to other types of lasers. The DF laser radiation spectrum falls into the so-called 'transparency window' of the atmosphere, as a result of which the sensing radius significantly increases. The electrophysical characteristics of the mixture of electric-discharge HF and DF lasers do not practically differ, the difference manifests itself in optical parameters – in mixtures containing a heavy isotope of hydrogen, the generation spectrum shifts to the long-wavelength region ($\lambda =$ 2.6 ÷ 3.1 μm for the HF laser and $\lambda = 3, 6 ÷ 4.1$ μm for the DF laser). The lasing spectrum of the non-chain DF lasers is limited in the long-wavelength range compared to the chemical chain-reaction DF lasers, the spectrum of which contains lasing lines from to 5 μm. Since there is also an atmospheric transparency window in the spectral range of 4.5–4.8 μm, obtaining high-power lasing in the spectral range of 4–5 μm is important for many special applications.

Today, sources of high-power lasing in the spectral region of 3-5 μm are in demand in various fields of science and technology – remote sensing of the atmosphere, medicine, special applications, etc. These applications require efficient laser sources with high pulse energy, pulsed and average power. Therefore, the development of the physical fundamentals of creating high-energy, effective non-chain HF(DF) lasers and laser complexes based on them, emitting in the spectral range of 2.6–5 μm, is an extremely important task.

The main goal of the book was the search for new principles of the formation of SSVD for creating high-energy non-chain HF(DF) lasers, as well as the creation of efficient lasers with high output energy and radiation power in the spectral region of 2.6–5 μm.

Since the understanding of the conditions for the formation of a homogeneous SSVD is of decisive importance when creating powerful electric-discharge non-chain HF(DF) lasers, the following research problems were posed and solved in this work.

Research objectives

1. To study the influence of the conditions of formation of SSVD, gas composition and the mode of energy input into the gas on the efficiency and radiation energy of non-chain HF(DF) lasers.

2. Conduct research on SSVD in mixtures of non-chain HF(DF), including a study of the stability and dynamics of the development of SSVD, as well as the effect of the gas composition and geometry of the discharge gap (DG) on its characteristics. On the basis of the conducted research, recommendations have been proposed for gas composition and for the method of obtaining SSVD in non-chain HF(DF) lasers.

3. Develop simple and reliable wide-aperture non-chain HF(DF) lasers and investigate their characteristics.

4. Investigate the possibilities of expanding the lasing spectrum of the non-chain HF(DF) lasers in order to create a powerful radiation source tunable in the spectral region of 3–5 µm.

The scientific novelty of the work consists in the study of the physical foundations of the formation of a scalable self-initiated volume discharge (SIVD) and the creation of high-energy wide-aperture laser systems based on it, as well as the creation and study of lasers based on optically excited using a non-chain HF(DF) laser crystal structures of ZnSe and ZnS doped with iron ions in the spectral region of 3.7–5 µm.

The scientific and practical significance of the results of this book lies in the fact that in it:

– A special form of a volumetric discharge, SIVD in SF_6 and mixtures on its basis was identified and investigated for he first time. It is shown that the formation of a SIVD in highly electrically negative gases occurs as a result of the processes of self-organization of dissipative structures, which are diffuse plasma channels. Preliminary ionization of the gas is not required to obtain the SIVD

– The effect of limiting the current density in the diffuse channel, causing the existence of such a form of the SSVD as the SIVD, has been experimentally discovered and investigated For the first time in SF6 and a number of other highly electronegative polyatomic gases, C_3F_8, C_2HCl_3, C_3H_7I,

It was shown that if there are small-scale inhomogeneities (~50 µm) on the cathode in SF_6 and a number of other highly electronegative polyatomic gases C_3F_8, C_2HCl_3, C_3H_7I, it is possible to obtain SIVD with uniform distribution of energy input in the DG volume, which makes it possible to create extremely simple and compact laser systems.

For the first time, a new approach to the problem of obtaining a homogeneous gas-discharge plasma in large volumes containing highly electronegative gases at medium pressures was proposed, which made it possible to increase the lasing energy of non-chain HF(DF) lasers initiated by SSVD more than 40 times.

The pulsed non-chain HF(DF) lasers with record power characteristics: W_{out} = 410 J at HF and W_{out} = 330 at DF (pulsed power P_{out} = 1.4 GW at HF and P_{out} = 1.1 GW at DF with an electrical efficiency of 4, 3% and 3.4% respectively), were constructed.

The temperature dependences of the critical reduced electric field strength $(E/N)_{cr}$ in undissociated SF_6 and SF_6:C_2H_6 mixtures in the temperature range T = 300–2400 K are obtained for the first time.

For the first time, high-power laser systems emitting ZnSe and ZnS crystals doped with a non-chain HF(DF) laser doped with non-chain HF(DF) laser in the spectral range of 4–5 μm were created. It is shown that the conversion efficiency of pump radiation in the ZnSe:Fe$_2$ + crystal structure at room temperature is ~50%.

The results of the work are of interest for understanding the conditions for obtaining volume discharges in various highly electronegative gases and can be used to create both high-power gas lasers and plasma-chemical reactors with large volumes.

The reliability of the results obtained is ensured by the high level of experimental technology, the creation of actually working devices; application of modern theoretical concepts and processing methods in the analysis of data; good agreement, obtained by different methods of results, as well as with available literature data.

The main provisions for the defense of the work:

1. In highly electronegative polyatomic gases: SF_6, C_3F_8, C_2HCl_3, C_3H_7I and mixtures based on them, the conductivity of the non-contracted plasma channel decreases with increasing electrical energy introduced into the channel. In SF_6 and mixtures based on it, a decrease in the conductivity of an uncontracted plasma channel is due to the joint action of two processes: the dissociation of polyatomic molecules by electron impact and electron–ion recombination.

2. In highly electronegative polyatomic gases: SF_6, C_3F_8, C_2HCl_3, C_3H_7I and mixtures based on them, a homogeneous volume discharge (even under highly heterogeneous distribution of the electric field in

the DG) is formed without using special pre-ionization devices due to the formation of a large number of diffuse plasma channels and non-stationary self-organizing processes of establishing the distribution of current density in them. The homogeneity of SIVD increases when using cathodes with a rough surface on which there are microscopic irregularities with characteristic dimensions of $50 \div 100$ μm.

3. Ignition of SIVD in working environments of non-chain HF(DF) laser allows creating non-chain HF(DF) lasers with an output energy of more than an order of magnitude greater than the level reached in the world before our research. The maximum radiation energy of a non-chain HF(DF) laser, initiated by a SIVD, increases linearly with the energy introduced into the working volume at energy inputs into the SIVD plasma $w_{in} = 0.5–2.5$ kJ $\cdot$ l^{-1} $\cdot$ atm^{-1}. The maximum efficiency of conversion of the electrical energy introduced into the plasma of a volume discharge into the radiation of a non-chain HF laser exceeds 4% at a pump energy of $W_{in} = 0.01 \div 10$ kJ.

4. The use of the non-chain electric-discharge HF laser for optical pumping of $ZnS:Fe^{2+}$ and $ZnSe:Fe^{2+}$ crystal structures allows the creation of high-energy laser systems that effectively emit in the spectral range 3.7–5 μm at room temperature: on a polycrystalline $ZnSe:Fe^{2+}$ structure with pulse energy more than 1.5 J, a pulse power of over 6 MW and a conversion efficiency of 48%; on the $ZnS:Fe^{2+}$ crystal structure with a pulse energy of 660 mJ and a conversion efficiency of over 26%.

5. In the temperature range $T = 300–2400$ K, the reduced electric field strength $(E/N)_{cr}$ in non-dissociated SF_6 and SF_6: $C_2H_6 = 5:1$ mixtures increases linearly with temperature. The increase in the reduced electric field strength $(E/N)_{cr}$ in undissociated SF_6 and C_2HCl_3 with increasing temperature is due to an increase in the rate of attachment of electrons to vibrationally excited SF_6 and C_2HCl_3 molecules.

6. Heating of SF_6 and gas mixtures based on it at $200 \div 2000$ K leads to the separation of and the formation of plasma instabilities in the discharge volume. The decrease in the rate of the electron–ion recombination process due to a local increase in temperature at the top of the plasma channel, which grows from the cathode spot (CS), leads to a contraction of the discharge in the working mixtures of non-chain HF(DF) laser.

Approbation of work. The main results of the work were reported and discussed at the seminar of the Department of Oscillations of the Institute of General Physics of the Russian Academy of Sciences and at 29 International and Russian conferences: (International Conference on Atomic and Molecular Pulsed Lasers (Tomsk, 1999 2001, 2007, 2015); International Conference on Plasma Physics and Plasma technologies (Minsk, 1997, 2000, 2003, 2006, 2009, 2015); Conference on the physics of gas discharge (Ryazan, 1996, 1998, 2000 and 2002); VIIIth International Conference on Gas Discharge and their Applications, GD-2000; International Symposium on High Power Laser Conference (1998, 2000, 2004); Int. Conf. High-Power Lasers in Energy Engineering, (Osaka, 1999); Conference on Laser Optics, (St-Petersburg, Russia, 2000, 2010, 2012); XXV International Conference on Phenomena in Ionized Gases, ICPIG-2001; International Conference on Lasers 96-99, Laser Interaction with International Symposium (LIMIS 2010), International (Zvenigorod) Conference on Plasma Physics and controlled thermonuclear fusion, (Zvenigorod, 2007, 2008, 2009, 2010, 2011, 2013, 2014).

Publications. Based on the materials of the research, 90 publications were prepared, of which 60 articles were published in publications defined by the Higher Attestation Commission of the Russian Federation, of which 51 articles were published in the leading peer-reviewed scientific journals included in the international bibliographic database WoS, 1 patent for the invention of the Russian Federation; 18 papers published in the works of national and international conferences.

The work was carried out in the Powerful Lasers Department of the Institute of Physics, RAS (before 2003) and in the Department of oscillations at the Institute of Physics, Physics, RAS (Laboratory of the physics of pulsed gas-discharge lasers). The research results were obtained with partial support of the RFBR grants: No. 05-08-33704a "Volumetric independent discharge in highly electronegative gases and its use in various electrophysical devices"; No 06-08-00568a "Critical electric field in vibrationally excited SF_6 and mixtures based on it"; No. 08-08-00242a "Volumetric independent discharge in highly electronegative gases under conditions of shock-wave disturbances caused by laser heating of the medium"; No. 09-02-00475-a "Study of the physics of the formation of working media of electrochemical non-chain HF(DF) lasers"; No. 15-02-06005-a "Investigation of the possibility of expanding the generation spectrum of non-chain HF(DF) laser."

Personal contribution of the author

The book is the result of the author's many years of work in the Laboratory of the Physics of Pulsed Gas-Discharge Lasers of the

Department of ML until 2003 and the Department of Oscillations of the A.M. Prokhorov Institute of General Physics the Russian Academy of Sciences (RAS) and is a synthesis of the works of the author, carried out jointly with the staff of the Institute of General Physics of the RAS. The collective nature of the experimental work led to the publication of the results obtained in collaboration with colleagues. The goals and objectives of the study were determined either by the author personally or with his participation. The author of this work was directly involved in the development of research methods, conducting experiments, creating calculation programs and processing the results obtained. Analysis, synthesis of the results and formulation of the conclusions of the work were carried out personally by the author.

All the main results of the work were obtained by the author personally, or with his direct participation. The main work was done in collaboration with Prof. V. Apollonov by K.N. Firsov, as well as other employees of the Institute of General Physics of the Russian Academy of Sciences (Kononov I.G., Marchenko V.M., Nefedov S.M., Oreshkin V.F., Podlesnykh S.V., Sayfulin A.V.) who participated and assisted in the conduct of individual studies. A part of experimental studies of a pulse-periodic (PP) mode of operation of non-chain HF(DF) lasers, was performed jointly with employees of the RFYaTs VNIIEF (S.D. Velikanov, S. Garanin, A. Domazhirov, N. Zaretsky), Kodola B.E., Komarov Yu.V., Sivachev A.A., Shchurov V.V., Yutkin I.M.); employees of the NGO FID-equipment (Efanov V.M., Efanov M.V., Yarin PM) and employees of the FSUE GNIILTS RF Raduga (Bulaev V.D., Gusev VS, Lysenko S .L. And Poznyshev A.N.). Works on studying the laser characteristics of the polycrystalline structures of ZnSe and ZnS doped with iron ions were performed together with the staff of the Institute of Chemical Biology and Chemical Sciences of the Russian Academy of Sciences (Gavrishchuk E.M., Ikonnikov V.B., Kotereva N.A., Mozhevitina E.N., Rodin S.A., Savin D.I., Timofeyeva N.A.), who provided samples for research and participated in the analysis and discussion of the results of the work. Separate works were performed jointly with the staff of NPO ALMAZ and Astrophysics (Ignatiev A.B., Feofilaktov V.A., Rogalin V.E.). Some studies of the physics of discharge in mixtures containing highly electronegative gases were carried out in a creative collaboration with A.A. Belevtsev, the employee of the Institute of High Temperatures of the Russian Academy of Sciences. In all cases of citing the results of research of other authors in the work references is provided to sources of information.

The structure and scope of the book. The book work consists of introduction, six chapters, conclusion and list of references. The total volume of the book is 220 pages, including 135 figures, 4 tables and a bibliography of 223 titles.

Main contents of the work

In the introduction, the relevance of the research topic is substantiated, the goal, tasks and main scientific statements of the work are formulated. The novelty and practical value of the results obtained in the book is noted. A summary of the contents of the chapters of the book will be given.

The first chapter contains a review of works devoted to the creation of high-energy non-chain electric-discharge HF(DF) lasers and the possibilities of expanding their lasing spectrum. Analyzed is the composition of the working environment, which achieved maximum efficiency. It is shown that the most effective donor of fluorine atoms is SF_6. Based on the literature data, the possibility of controlling the lasing spectrum of an HF(DF) laser and the possibility of obtaining lasing in a wider spectral range are analyzed. Methods for obtaining the self-sustained volume discharge (SSVD) and electrical circuits for their implementation have been studied, which made it possible to achieve maximum energy characteristics of non-chain HF(DF) lasers. Special attention was paid to the possibility of scaling installations based on these methods. Based on the analysis, several features of SSVD formation in non-chain HF lasers were revealed, and it was concluded that when creating high-energy lasers, traditional SSVD methods developed mainly for CO_2-pumped transverse discharge lasers have extremely limited application in the case of electric-discharge HF(DF) lasers with large apertures.

The *second chapter* presents the results of the study of SSVD in gas mixtures based on SF_6, from which it follows that the discharge has a fundamentally jet structure (represents a set of overlapping diffuse channels), and the combination of the observed features of the development of SSVD is due to the existence of processes that limit the current in a separate channel (the effect of current density). Numerical models are presented that allow the calculation of the SSVD characteristics in the working mixtures of a non-chain

HF(DF) laser, as well as qualitatively investigate the dynamics of the formation of the SSVD.

The *third chapter* of the book presents the results of studies of the plasma of a single diffuse channel in SF_6 and SF_6 mixtures with hydrocarbons. Mechanisms are investigated that allow obtaining SSVD in working mixtures of the HF(DF) laser without using special pre-ionization devices. The general patterns of formation of SSVD in SF_6 and mixtures based on it are discussed.

The *fourth chapter* presents the results of studies of the influence of pulsed laser heating of gas on the characteristics of SSVD in gas mixtures based on SF_6 and analyzes the development of distinctive instability in active media of electric-discharge non-chain HF(DF) lasers due to electron detachment from negative ions.

The *fifth chapter* of the book presents the results of studies of the characteristics of non-chain HF(DF) lasers initiated by SSVD. The problems of creating non-chain HF(DF) lasers of a kilojoule level and *P–P* lasers operating with a high pulse repetition rate are discussed. The results of the study of the possibility of monitoring the wavefront (HF) of radiation of non-chain HF(DF) lasers using the methods of Talbot interferometry are presented.

The *sixth chapter* presents the results of studies performed with the use of the created non-chain HF(DF) lasers. The concept of creating high-power laser systems emitting in the spectral region of 3–5 μm based on the ZnSe and ZnS crystalline structures. doped with iron ions when pumped by a non-chain HF laser. is presented.

In conclusion, the main results of the work are listed.

The book ends with a bibliographic list of the cited studies.

High-energy HF (DF) lasers with non-chain chemical reaction (literature review)

In this chapter, we analyze the methods and approaches used in the creation of pulsed and P–P non-chain HF (DF) lasers with high radiation energy.

1.1. Chemical HF (DF) lasers. (The principle of operation and general characteristics)

With the discovery of chemical lasers (CL) high hopes were pinned on them. They were connected with the possibilities of their use in various fields, mainly in the field of military applications, in power engineering, in the creation of laser engines, etc., i.e. in those areas where laser beams with low divergence and extremely high energy characteristics are necessary [1–4]. A distinctive feature of the CL is the possibility of obtaining laser energy with an efficiency higher than 100%, i.e. the emitted energy may be several times higher than the energy expended to initiate a chemical reaction. A classic example of such a laser is a chain HF(DF) laser, in which lasing in the mid-IR range occurs on the vibrational–rotational transitions of the HF(DF) molecule, which is formed when fluorine gas interacts with hydrogen (deuterium). The characteristic reactions of the processes are as follows:

$$H + F_2 \Rightarrow HF(v) + F; \quad F + H_2 \Rightarrow HF(v) + D;$$
$$HF(v) + hv \rightarrow HF(v\text{-}1) + 2hv \quad \lambda = 2.5 \div 3.1 \ \mu m \quad (1.1)$$
$$D + F_2 \Rightarrow DF(v) + F; \quad F + D_2 \Rightarrow DF(v) + D;$$
$$DF(v) + hv \rightarrow DF(v\text{-}1) + 2hv \quad \lambda = 3.5 \div 4.8 \ \mu m \quad (1.2)$$

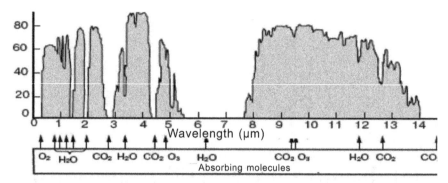

Fig. 1.1. Transmission of the atmospheric surface layer (15°C, humidity 40%). Shaded areas are atmospheric transparency windows [5, 6].

The energy characteristics of The CL are record-breaking among the whole class of gas lasers. An additional factor that stimulated the development of HF(DF) laser research was that the DF laser radiation is little absorbed by atmospheric gases (falls into the atmospheric transparency window) (Fig. 1.1). Figure 1.1 shows the dependence of the atmospheric transmission on the radiation wavelength [5].

Due to the high interest in the DF laser from the military-industrial complex [6, 7], by the beginning of this century, laser systems of the megawatt level had already been created, including in the mobile version [8]. The main drawback of the chain CL, greatly limiting their field of application, is the high toxicity and explosiveness of the starting components. It is because of this that it is possible to work with chain HF(DF) lasers only on specially equipped sites. Another significant drawback of the chain CLs is the difficulty of implementing the IP mode, especially when a number of applications require work with a high pulse repetition rate (~100 Hz and more). At the same time, lasers on non-chain reactions lack the disadvantages listed above [3, 9]. The efficiency and energy parameters of non-chain HF(DF) lasers are not as high as those of chain analogs, but the lasing spectrum also lies in the average IR range ($\lambda = 2.6 \div 3$ μm for the HF laser and $\lambda = 3.6 \div 4.1$ μm for the DF laser), while they are safer and more convenient to use. The first papers describing the generation of non-chain lasers based on vibrational-rotational transitions of HF and DF molecules appeared in 1967 [10, 11], i.e. long before the CL were created on chain reactions. Over the next 50 years, interest in the non-chain HF(DF) lasers was unstable. Periods of a sharp increase in the number of publications devoted to HF(DF) laser studies (or their applications) alternated with periods of decline, when the activity of research groups in this area decreased markedly. The uneven activity of scientific groups working in the field of the non-chain HF(DF) CLs is due to several reasons. First, the progress in the development of such lasers required for their decision a deeper understanding of the physical processes occurring in the gas-discharge plasma, the application of new approaches, the development of technologies or the development of a new elemental base. The second reason is associated with progress in the field of creating solid-state lasers, which gave hope for obtaining comparable energy characteristics, including in the spectral region of 2.7–4.5 μm. This is the reason for the decrease in the level of funding for other areas of research. However, even today, in the spectral region of 2.6–4.2 μm, the energy characteristics — pulse energy, pulsed and average power, of the non-chain HF(DF) lasers are still significantly superior to solid-state lasers.

The principle of the non-chain HF(DF) CL is based on the fact that in a laser working medium, which is a stable mixture of fluorine and hydrogen-containing (deuterium-containing) gases, a non-chain reaction is initiated by the input of energy from an external source. Molecules of the starting materials dissociate to form chemically active centres (fluorine atoms), the reactivity of which is many times greater than the reactivity of the starting materials. As a result of the production of fluorine atoms in the working volume, it becomes possible for a chemical reaction to form HF(DF) molecules in a vibrationally excited state [3, 9]. In a non-chain laser, the active centres form only as a result of the external energy source, and after the cessation of energy supply, the chemical reaction quickly fades [12]. In a simplified form, these processes can be written as follows:

1) $W + AF \rightarrow A + F$ – dissociation of the fluorine-containing component AF due to the supply from an external energy source W.

2) $F + RH (RD) \rightarrow HF^* (DF^*) + R$ – interaction of F atoms with a hydrogen or deuterium-containing component RH (RD), as a result of which a vibrationally excited molecule HF (n) (or DF (n) is formed), where n is the vibrational quantum number ($n > 0$).

3) $hv + HF(n) \rightarrow HF(n-1) + 2hv$ – radiation of a quantum of light.

As a working medium of a non-chain laser, a mixture of SF_6 with hydrogen (deuterium) is mainly used. SF_6 is a non-toxic, chemically stable compound that is used in fire fighting as a gas flame arrester, and in high-voltage technology as a gas electrical insulation [13]. Reactions in the $SF_6:H_2$ (D_2) gas mixture become possible only after

4

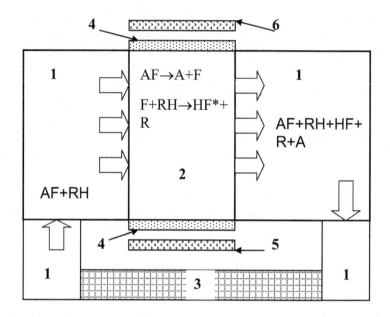

Fig. 1.2. P–P scheme of non-chain HF(DF) laser: 1 – chamber; 2 – laser zone, where using an external energy source cause the dissociation of the fluorine-containing component; 3 – filter for binding HF (DF); 4 – windows for outputting radiation to the resonator; 5, 6 – resonator mirrors.

SF_6 molecules dissociate with the formation of fluorine atoms due to an external energy source.

A schematic diagram of the P–P non-chain HF (DF) laser is shown in Fig.1.2. The flow of non-reacting gases is pumped through laser zone 2, where a non-chain chemical reaction is initiated from an external energy source. The reaction in this mixture is possible only while the fluorine atoms are being generated. After the laser zone, the mixture is pumped through filter 3, on which HF(DF) molecules and SF_6 dissociation products are deposited, and the mixture again enters the laser zone. The degree of dissociation in the laser zone, as a rule, does not exceed 5%; therefore, the gas consumption for one cycle is insignificant (in real lasers, usually the volume of the laser chamber significantly exceeds the volume of the laser zone) [9]. During long-term operation with a high pulse repetition rate in a non-chain HF(DF) laser, a continuous supply of a fresh mixture is carried out to compensate for the reacted components.

The non-chain HF(DF) lasers are characterized by large values of the gain, while lasing occurs according to a cascade mechanism,

Table 1.1. The energy effect of the chemical reaction of the interaction of fluorine atoms with various hydrogen donors RH (deuterium RD) and the fraction of this energy (η_x), which is used to excite vibrational levels of the HF (DF) molecule

F+RH(RD)=HF(DF)+R RH(RD)	Q, eV	Q, kcal/mol	Chemical efficiency η_x,
H$_2$	1.43	32.9	0.71
D$_2$	1.4	21.7	0.7
HI	1.95	44.8	0.95
HBr	2.11	48.5	0.24

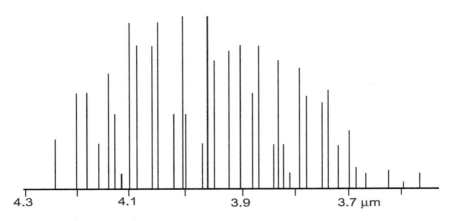

Fig. 1.3. Relative intensity of lasing lines of a pulsed DF laser [15].

i.e. the lower laser level of the previous transition is the upper for the next [14]. For example, transitions in the vibrational bands $n = 3 \rightarrow n = 2$, $n = 2 \rightarrow n = 1$, $n = 1 \rightarrow n = 0$ are observed in an electric-discharge HF laser based on SF$_6$:H$_2$ mixtures [3]. In this regard, the laser has a rather complex spectrum of output radiation, which is a large number of closely spaced lines corresponding to different rotational numbers of vibrational transitions [2, 10]. Figure 1.3 shows a typical lasing spectrum of a chemical DF laser [15]. (The spectrum of a non-chain laser is usually poorer [3], but depending on the particular implementation in the spectral region up to 4.1 μm, it is similar to the lasing spectrum shown in Fig. 1.3). When using a selective cavity mirror or introducing a selective absorber inside the cavity, it is possible to control the spectral composition of non-chain CLs quite efficiently, highlighting certain vibrational-rotational lines [16].

The amount of energy concentrated in the vibrational degrees of freedom of the HF(DF) molecules is largely determined by what substance was used as a hydrogen (deuterium) donor [3, 17, 18]. Table 1.1 presents data on the amount of energy released during the chemical exothermic reaction (Q), as well as its fraction (η_x), which is used for vibrational excitation of the HF(DF) molecule for various hydrogen donors (deuterium). The value of η_x is at the same time the limiting value of the chemical efficiency of an HF(DF) laser in mixtures containing the RH(RD) molecules. It should be noted that Q and η_x do not depend on the specific form of the fluorine-containing compound AF, since the AF compounds do not directly interact with RH, and a false assumption may arise that the lasing spectrum of the HF(DF) laser also does not depend on which fluorine-containing compound the substance is used. However, this is not the case. The fact is that AF dissociation products are formed not only in the ground state, but also in the excited state, and thus carry some excess energy, the magnitude of which, with the chosen initiation method, depends on AF [3]. The choice of AF also has a direct effect on the overall efficiency of the chemical laser, since the main energy consumption in non-chain HF (DF) lasers is the dissociation of AF fluoride molecules to form F atoms. Most studies have been carried out on lasers in which fluoride dissociation occurred due to electron impact, in the literature this type of lasers is also called electrochemical [9, 21]. It should be noted that the difference in the energy consumption for the dissociation of AF molecules leads to different absolute values of the gain of the active medium of non-chain HF(DF) lasers, which also affects the lasing spectrum [18, 19].

In [20], it was shown that the problem of creating an effective non-chain HF(DF) laser reduces, firstly, to the choice of suitable fluorine and hydrogen-containing gases, in which, with minimal initiation energy, it is possible to obtain the maximum output energy of the laser W_{out}, and second, to the choice of an effective method of initiating a non-chain chemical reaction.

The method based on the dissociation of fluorine-containing molecules by electron impact, carried out in a self-sustained volume discharge (SSVD), has the greatest advantages (low energy costs for the formation of fluorine atoms, the possibility of exciting large volumes of the medium, etc.) over all other methods of initiating non-chain chemical reactions [20]. In contrast to photolysis, this method is also more versatile, since it allows the use of a much wider range of fluorine-containing compounds, and in contrast to the use of an electron beam, it does not require sophisticated protection from X-rays and has higher characteristics in terms of structural reliability. At present, among various gas lasers, it is electric discharge systems that have found the greatest application. Electric-discharge non-chain HF(DF) lasers are used in medicine [22–25], for optical pumping of crystals [26], in laser chemistry [27–29], in military science [7], for recording holograms [30] and in environmental monitoring [5, 31, 32]. In this regard, we consider electric discharge HF(DF) lasers in more detail.

1.2. HF (DF) lasers with the initiation of non-chain chemical reaction by electric discharge

The first lasing on the HF in an electric discharge was obtained by Deutsch in 1967 [10]. After that, the main effort of researchers was focused on finding the most appropriate fluorine and hydrogen-containing compounds and improving the methods for obtaining the discharge [17, 33–35]. In the first works, schemes with a longitudinal electric discharge were used (the highest characteristics of the HF laser with this method of excitation were obtained in [36]), but the possibility of increasing the working volume and laser energy was limited by the need to supply extremely high voltages to the electrodes. Therefore, almost immediately after the appearance of the first publications on the creation of a CO_2 laser with pumped transverse SSVD, the methods developed for this purpose began to be used to initiate a chemical reaction in the non-chain HF(DF) lasers [37–42]. It is this type of electric discharge that allows more energy to be introduced into the gas and to excite large volumes of the active medium [43, 44]. Independently of one another by many experimental groups, it was found that the best parameters of a non-chain HF laser are realized when SSVD is ignited in the $SF_6{:}H_2$ or $SF_6{:}C_2H_6$ working mixtures.

It should be clarified why the SSVD in gas mixtures containing fluorine atoms as SF_6 as a donor turned out to be the most effective way to initiate a chemical reaction in the non-chain HF(DF) lasers. When an electric current flows in a gas, the main parameter that determines such parameters as the average electron energy, ionization constants, adherence, etc., is the parameter E/N, where E is the

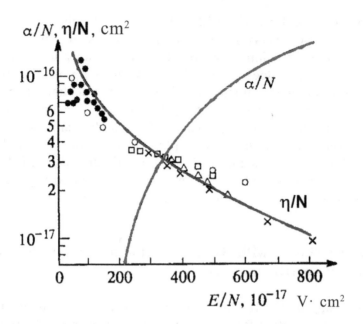

Fig. 1.4. Dependence of the ionization factor and the attachment factor of electrons in SF_6 on E/N parameter (at room temperature).

magnitude of the electric field strength and N is the number of gas particles per unit volume [45]. Figure 1.4 shows the dependences of the ionization (α) and attachment (η) coefficients on E/N in SF_6 [13]. The point of intersection of these curves determines the reduced critical field $(E/N)_{cr}$, which remains approximately constant during the SSVD discharge pulse [46, 48]. The value of $(E/N)_{cr}$ sets the average electron energy in the discharge and the rate constant for various plasma-chemical processes [45]. It is noteworthy that due to the strong electronegativity of SF_6, electron multiplication occurs at a very high value of the parameter $(E/N)_{cr}$ and the average electron energy in an independent discharge is almost 10 eV, with more than 80% of the energy going to dissociate SF_6 with the formation of fluorine atoms [49]. The value of the parameter $(E/N)_{cr} = 360$ Td, which is set in the discharge, is close to the optimum for SF_6 dissociation (1 $Td = 10^{-17}$ V cm²). As a result, the energy cost of the formation of fluorine atoms in SF_6 is the lowest among stable and non-toxic fluorides — $E_F \approx 4$ eV [50, 51]. Table 1.2 shows the values of the energy cost of a fluorine atom during electron beam dissociation of various molecules. Note that in some respects a non-chain HF laser with electric discharge initiation turned out to be

Table 1.2. The cost of the formation of fluoride atoms when dissociated by electron impact of various fluoride-containing compounds

F+RH(RD)=HF(DF)+R RH(RD)	Q, eV	Q, kcal/mol	Chemical efficiency η_x,
H_2	1.43	32.9	0.71
D_2	1.4	21.7	0.7
HI	1.95	44.8	0.95
HBr	2.11	48.5	0.24
CH_4	1.5	34.5	0.6
C_2H_6	1.7	39	0.62
C_4H_{10}	1.66	38	0.56
CH_3Cl	1.61	37	0.68
$CHCl_3$	1.83	42	0.37
CH_3Br	1.56	35.9	0.67
CH_2Cl_2	1.66	38	0.51
$C_2H_3F_3$	1.46	33.5	0.67
C_6H_{12}	1.86	42.9	0.53

much simpler than other lasers excited by a self-sustained volume discharge. For example, for a CO_2 laser, the value of $(E/N)_{cr}$, which is established in the self-sustained discharge, is higher than the optimum, and vice versa in nitrogen, and we have to resort to various tricks to increase the efficiency of these lasers [52]. In working mixtures of the non-chain HF (DF) laser, nothing of the kind needs to be done – it is only necessary to obtain a uniform discharge in the laser working mixture.

As the technique of producing SSVDs was improved, many different schemes of electric-discharge non-chain HF lasers were proposed [2, 20]. In fact, all these systems reproduced the electrical circuits and design of discharge chambers, which were previously developed for CO_2 and excimer lasers, without taking into account the specifics of the gas mixtures used in HF laser working environments. Therefore, the radiation energy of the most efficient non-chain electric-discharge HF laser until 1996 (beginning of this work) did not exceed 11 J [53], and from 1970 to 1996 the radiation energy of CO_2 lasers pumped by an electric discharge increased by more than three orders of magnitude, and by the beginning of this century, it reached ~5 kJ [54]. Excimer lasers were also developed rapidly [55].

1.3. The problem of increasing the energy characteristics of non-chain HF lasers

Low progress in increasing the output characteristics of non-chain electric-discharge HF lasers was associated by most researchers with the difficulty of obtaining (in large discharge volumes) SSVD in SF_6-based gas mixtures. Indeed, the traditional approach to the ignition of SSVD in dense gases, which was formed by the 90s of the last century and became classical already [43-48, 55], involves the creation of volumetric gas pre-ionization (UV, soft X-rays) with subsequent application to the discharge gap (DG) with a homogeneous electric field of high-voltage pulse [43]. In non-chain HF(DF) laser working mixtures, the ability to meet these requirements is limited by the strong electronegativity of SF_6. In addition, the simplest sources of UV radiation in this case are not effective due to the strong absorption of UV radiation in SF_6 [56]. Figure 1.5 shos whe dependence of the SF_6 photoabsorption cross section on the radiation quantum energy [56]. It is easy to see that with an SF_6 pressure of $p = 60$ mm Hg the characteristic photon path length with an energy of about ~10 eV (the ionization potential of 'heavy' hydrocarbons) is only $l_\varphi = \dfrac{1}{\sigma_\varphi N_{SF_6}} \sim 5$ mm, which is clearly not enough to initiate SSVD in wide-aperture lasers. The creation of soft X-ray sources with an emitting area in excess of 100 cm² is an independent problem [55, 57, 58]. In addition, SF_6 is a highly electronegative gas, which causes large losses of free electrons in the process of sticking and their rapid disappearance in the discharge gap. It is enough to note that already at $E/p \approx 0.9\,(E/p)_{cr}$ (E is the electric field strength in the discharge gap, p is the pressure of SF_6) and at a pressure of SF_6 $p = 60$ mm Hg, the electron lifetime is $\tau_l = 1/(\eta \times v_e) \approx 1$ ns. (Here η is the sticking coefficient, v_e is the electron drift velocity).

The presence of such a powerful channel for the death of free electrons places high demands on the power of the pre-ionization source and almost nullifies the possibility of fulfilling the condition of creating initial electrons with a concentration $n_e \sim 10^8$ cm^{-3} in the entire discharge volume. Additional difficulties with increasing the aperture and volume of the active medium of non-chain lasers arise due to the need for special profiling of the electrodes to ensure the uniformity of the electric field in the DF, since this leads to an

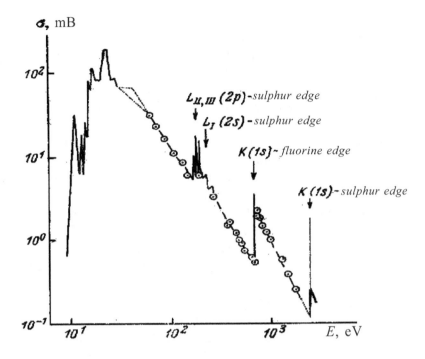

Fig. 1.5. The dependence of the cross section of the photoabsorption of SF$_6$ on the energy of a soft X-ray quantum [56].

increase in the size of the laser and the inductance of the discharge circuit with extremely limited duration of stable combustion of OCP in gas mixtures containing SF6 [39, 59].

The parameters of a non-chain HF laser initiated by an SSVD with an energy of ~400 J were estimated on the basis of literature data [20]. The choice of the composition of the working environment was quite obvious – SF$_6$ should be used as a donor of fluorine atoms. The reasons for this choice are, firstly, in the convenience of operation (SF$_6$ is a non-toxic and non-flammable gas) and, secondly, in the low cost of formation of fluorine atoms, $q_F \approx 4$ eV [50] (see Table 1.2). Indeed, as established in [35], the maximum efficiency of an electric-discharge non-chain HF laser is achieved when using SF$_6$ atoms as a donor of fluorine atoms. The choice of the second necessary component of the medium, namely, the donor of hydrogen atoms RH, is not so unambiguous, since this component determines not only the chemical efficiency, but also to a large extent the SSVD characteristics [3]. Since this issue will be discussed in detail in the next section, here we only note that most often H$_2$ or hydrocarbons (C$_2$H$_6$ and C$_3$H$_8$) are used as RH. Regarding the specific characteristics of non-chain HF(DF) lasers initiated by the SSVD, one important note needs to be made: in the literature one can find reports on the specific energy output $W_{sp} \sim 15 \div 20$ J/l [40], however, such high parameters were achieved only in installations with very small active volume (V), which actually did not exceed a few cubic centimeters. Numerous attempts by researchers to achieve the same W_{sp} values on units with $V \sim 1$ l were unsuccessful. Apparently, there is a common scaling problem for electric-discharge gas lasers. Therefore, when estimating the laser parameters, it is necessary to take into account not the limiting values of the specific characteristics of the laser, obtained in experiments on installations with a small active volume, but some average values that will be approximately preserved with increasing active volume. Analysis of the literature [32–42] shows that the average values of the specific energy output W_{sp} and the total electrical efficiency in SF_6:H_2 mixture lasers (hydrocarbons) are $W_{sp} \approx 5$ J/l and $\eta_T \approx 3\%$, respectively. The exceptions are quite compact systems [40, 60] with a working volume not exceeding 60 cm^3. Since such systems do not scale, the estimates should focus on more realistic average results. As a result, we obtain the following estimated parameters: a discharge volume of $V \approx 80$ l, an aperture of ~28 cm, and the energy input to the plasma of the SSVD $W_{in} \approx 100$ J/l. The problem of creating CO_2 lasers with a similar aperture and the same level of energy input was solved in the late 80s of the last century [54, 61]. However, attempts to increase the discharge volume and, accordingly, the radiation energy of non-chain HF(DF) lasers were far from being so successful as they were associated with low SSVD stability in the working mixtures of such lasers [62]. It is obvious that without solving the problem of the formation of a stable volume discharge in large discharge gaps, it is not possible to increase the energy of a non-chain laser. In this connection, let us pay attention to works in which the conditions for obtaining SSVD in the working mixtures of the HF(DF) laser differed from those usually considered.

1.4. Search for methods for SSVD formation in the working mixtures of HF(DF) laser

An interesting result was obtained in [27, 37, 63], the authors of which found that the electrode system used by them makes it possible to ignite a uniform SSVD at a working mixture pressure of up to

100 mm Hg without preionization. It is possible to note that it can be used in the workplace. without preionization. At the same time, at a pressure of 30–50 mm Hg, the effect of UV preionization on the generation level was weak [27], but preionization made it possible to stabilize the spread of laser energy from pulse to pulse (changes in the output energy from pulse to pulse during UV preionization were 1%, without preionization ~20 %). Taking into account the fact that the creation of initial electrons in working mixtures of an HF (DF) laser is a complex problem, and the rejection of a preionization source can significantly simplify the laser design, let us consider the electrode system [27, 63].

In [27, 63] the authors used a single crystal of germanium with with a specific resistivity of ρ =25 Ohm·cm and a mesh brass cathode through which UV preionization was carried out. To ensure a uniform field in the DG, the anode had a Chang profile. The laser operated on an SF_6–C_3H_8 mixture (technical propane–butane mixture with a C_3H_8 content of ≈60%; C_4H_{10}≈40%) and had high specific energy characteristics. Unfortunately, the authors of [27, 63] did not carry out any analysis of the reasons that make it possible to obtain SSVD at sufficiently high energy inputs W_{in}~100 J /l without preionization. As the main reason for the effect the authors considered the use of an anode of single-crystal germanium. A positive role in increasing the uniformity and stability of SSVD when using electrodes made of semiconducting materials was also reported in other publications [27].

However, an analysis of the literature on electric-discharge HF (DF) lasers allows one to doubt that the main reason that makes it possible to obtain SSVD without preionization is the use of a semiconducting anode with volume resistance. So, earlier it was possible to obtain SSVD without preionization in He:SF_6:C_3H_8 = 300–90015:1 mixtures at a total pressure of up to 600 mm Hg, reported in [29]. Installations [37] and [63] have close output energy characteristics W_{out}~400 mJ, power P_{out}~15 MW, however, with the same energy input W_{in}~100 J/l, the specific characteristics [63] are about 1.5 times better. The setup execution scheme [37] is similar to the scheme in [27, 63], only in [37] the discharge was ignited between the Al cathode made along the Rogowski profile and the mesh anode through which the discharge gap was illuminated by UV. It was also noted in [37] that sandblasting the cathode surface improves the structure of the discharge, while it is especially noted that the presence of C_3H_8 in the working mixture was a prerequisite for ignition of the SSVD, without which it was not possible to obtain an SSVD without a spark even in the presence of preionization . An improvement in the quality of the discharge when using a cathode with a rough surface in [37] was associated with an increase in the photoemissive capacity of the cathode (due to an increase in the effective area). The reasons for the influence of C_3H_8 on the improvement of the discharge were indicated: low, ~11 eV ionization potential, and a large photoionization cross section. However, no specific mechanisms for this influence were proposed, so we will consider this issue in more detail.

It is known [64] that the addition of certain substances with a low ionization potential to the CO_2 working media of lasers makes it possible to increase the stability of SSVDs by reducing the electron energy due to a change in the electron energy distribution function and, as a consequence, suppressing the development of instabilities due to stepwise ionization of nitrogen. It could be assumed that the effect of C_3H_8 (and other hydrocarbons) on the stability of SSVD in working mixtures of an HF laser has a similar nature, i.e. a change in the electron energy distribution function, but no special studies of this issue were carried out until 1996. As regards the relatively high value of the photoionization cross section of C_2H_6 (C_3H_8, etc.), it should be noted that with strong UV illumination and small distances from the illumination source to the active volume, this fact allows to increase the initial electron concentration. Apparently, this explains the improvement in the SSVD structure noted in [40] and the increase in the lasing energy of a the non-chain HF laser when H_2 is replaced by C_2H_6. In general, it is quite obvious that the use of compounds with a low ionization potential and large photoionization cross sections can reduce the requirements for the preionization source in small-sized installations, but it does not allow solving the problem of scaling the discharge volume in principle. In this regard, it is necessary to pay closer attention to the devices that were used to ignite the SSVD in [27, 63 and 37]. In these systems, the sources of UV illumination had a separate power supply, and when the authors of [27, 37, 63] say that the discharge was obtained without preionization, this only means that the power supply circuit of the backlight was not turned on. However, no special experiments with shielding of the spark gaps of the UV illumination source were performed in these studies. At the same time, it is well known that when the switches of the main discharge generator are turned on, spark gaps of the UV illumination source in such schemes can break through due to the induced potential difference on them. An indirect confirmation of the above is a relatively small scatter in the output energy. If the backlight of the main gap were completely absent, then in the case of a small cathode area in [27, 37, 63] and low overvoltages on the discharge gap, there should have been a significant spread in the delay time of the breakdown of the gap, and the spread in the values of the output energy would be much higher. However, in [27, 37, 63] there is generally no mention of the instability of the breakdown of the main gap under conditions when the source of preionization was not turned on.

Obviously, neither the spontaneous breakdown of some spark gaps in the preionization scheme, nor the corona from the surface of the electrodes can provide the necessary, according to traditional concepts, initial electron concentration at the level of $n_0 \sim 10^6$ cm^{-3}. It is noteworthy that the analysis of other works also points to the weak role of preionization in the formation of SSVD in the non-chain lasers. So, for example, in [41] and [53] approximately the same output energy and laser efficiency were obtained. In [41], where a series of metal pins connected to a common bus via resistances (resistive isolation) served as the cathode of the discharge gap, there was no preionization in the traditional representation (coronation from pins, but this is a very weak source), and in [53] it was carried out by a high-current a creeping discharge, in the emission spectrum of which not only UV radiation was present, but also soft X-ray radiation. Moreover, in [41] the electrical efficiency (3.8%) turned out to be even higher than in [53] (2%). In this regard, it is necessary to especially note the results obtained by the authors of [65, 66]. In [66], the high energy characteristics of an HF laser were achieved using a setup in which a barrier discharge distributed over the cathode surface was used to obtain SSVD. Moreover, in [66], the voltage on the discharge gap having a rather high edge amplification of the electric field was applied with a short front, which, in principle, did not allow uniform filling of the discharge gap by electrons from the plasma created at the cathode using a barrier discharge. We also note that the radiation intensity of the barrier discharge in SF$_6$-based gas mixtures is insufficient to create an initial electron concentration in the discharge gap even at the level $n_0 \sim 10^6$ cm^{-3}, which is necessary according to traditional ideas [52, 55].

Looking ahead, it should be noted that when conducting prospecting studies of the possibility of increasing the energy of a non-chain electric discharge HF laser at the end of the 90s of the last century, our research team found that in mixtures of SF_6 with hydrocarbons it is possible to obtain a volume discharge without preionization. Having realized the possible prospects for using this effect in lasers, our group began systematic studies of SSVD in gas mixtures based on SF_6. Subsequently, the shape of the discharge, which is realized in mixtures based on SF_6, was called a self-initiating volume discharge [68, 69] (see Chapter 2 of this dissertation) and the main task of the author in subsequent years was to study the physics of this peculiar form of volume discharge.

1.5. P–P non-chain HF (DF) lasers with a high pulse repetition rate

The physical aspects of the implementation of the P–P mode in non-chain electric-discharge HF(DF) lasers are considered in detail in Ref. [70], therefore, we present only the main results. Figure 1.6a shows a diagram of a laser setup in which the authors of [71] demonstrated the possibility of operation of a non-chain closed-circuit HF(DF) electric discharge laser with a repetition rate of up to 3 kHz. To obtain the SSVD in the discharge gap of $5 \times 5 \times 150$ mm^3, preionization with UV radiation from a number of spark sources located along the surface of the electrodes was used. The duration of the discharge pulse in this laser did not exceed 100 ns. The laser worked on mixtures of $SF_6 : H_2 = 10 : 1$.

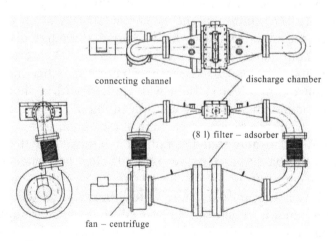

connecting channel   discharge chamber

(8 l) filter – adsorber

fan – centrifuge

a

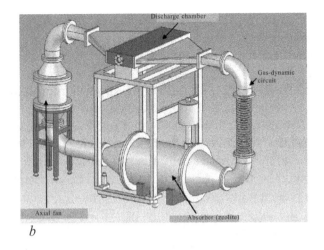

b

Fig.1.6. Scheme of a non-chain electrochemical HF (DF) closed-loop laser: a) according to [71]; b) according to [72].

The results of the authors' studies [71] were the following observations:

1) The characteristics of the laser (output energy, the maximum possible pulse repetition rate at which the energy in the pulse does not decrease) strongly depended on the power of the preionization source. It was also noted that with an increase in the spark gap (due to the burning of the preionizer electrodes), the maximum frequency of the laser operation increases.

2) The speed of pumping working gas through the discharge gap should provide more than five-fold gas change in the working volume (in the article, the speed was stated that the flow rate provided an 8-fold gas change).

3) The consumption of components in the mixture under the P–P mode of laser operation was as follows: 2 H_2 molecules were consumed per SF_6 molecule.

4) Sorbents based on zeolite 5A, which were used for the adsorption of HF (DF), also absorb significant amounts of SF_6.

5) The adsorber is the largest element of the laser, through which it is necessary to purge the gas flow, and it should ensure complete absorption of the products of plasma-chemical reactions.

6) The main limiting factor limiting the laser operation time was the low survivability of the preionizer. The tungsten needle electrodes used for spark UV preionization quickly burnt out. A preionizer, whose electrodes are made of silicon, served for a much longer

time; the discharge from these semiconductor electrodes burned in quasi-volume form.

7) A corona discharge semiconductor preionizer provided the best laser parameters.

Much later, in Ref. [72], a similar scheme was reproduced (Fig. 1.6b), but they were able to significantly improve the stability of the P–P discharge when ethane was added to SF_6. Hydrocarbons not only increased the stability of the discharge, levelling the strong dependence of the laser characteristics on preionization, but also allowed to increase the laser operating time (i.e., the number of shots without reducing energy in a closed-loop mode). True, it should be noted that the pulse repetition rate in [72] was almost an order of magnitude lower than in [71], since in [71] there was a more powerful centrifugal fan, and Chinese researchers used a block of axial fans.

Many researchers working with non-chain P–P HF(DF) lasers (this is especially evident in low-aperture installations operating with a high pulse repetition rate) faced a dilemma: on the one hand, to stabilize laser characteristics at high frequencies, it is necessary to increase the preionizer power, and on the other hand, an increase in the power of the spark preionizer leads to rapid degradation of the working mixture and the burning of the preionizer electrodes. In this connection, the results of [73, 74] are noteworthy, in which the authors used a quasi-volume discharge for preionization, which develops from semiconductor electrodes made of SiC. It should also be noted that a detailed analysis of the design of the discharge chamber and the location of the UV preionizer in [71] allows us to note that the improvement in laser characteristics always correlated with an increase in cathode illumination. When the length of the arc channel from the preionizers was increased or a quasi-volume discharge was used, such a geometry of the preionizer source provided a significantly larger flux of UV radiation to the cathode. In general, we have to admit that many authors point to the degradation of the preionizer electrodes as the main factor that reduces the working time of a closed-loop laser. Therefore, the solution to the problem of increasing the life of the preionizer in a closed-loop laser is very important. Obviously, the possibility of completely abandoning the preionization system could greatly simplify the design and increase the life of the non-chain HF(DF) laser initiated by the SSVD.

When operating a non-chain HF(DF) laser in the P–P mode with a high pulse repetition rate, cooling of the working mixture is also an important factor. The release of additional thermal energy from preionization devices and from a gas stream that slows down when it encounters obstacles in its path is a negative factor. Many authors have shown that increasing the gas temperature in the working zone leads to a decrease in the output energy and laser efficiency, and, conversely, cooling the mixture leads to an increase in the output energy and the laser spectrum becomes richer [3, 38].

It should be noted that, in contrast to pulsed lasers, when implementing the P–P mode, the stability of a volume discharge has much higher requirements. A certain margin of stability is required, since in the discharge gap under conditions of a powerful gas flow and periodic disturbances from previous discharges, inhomogeneities in the density and temperature of the gas can occur. The best characteristics for solving several problems of an P–P laser at once were shown by the approach proposed in [75, 76]. In these works, it was shown that the use of a blade system of electrodes with inductive stabilization allows for high gas flow rates through RP and high discharge stability. Figure 1.7 shows a compact HF(DF) closed-loop laser circuit [78]. Using the approach described above, lasers were created with a pulse repetition rate of more than 3 kHz, as well as a laser with an average power of more than 400 W [77].

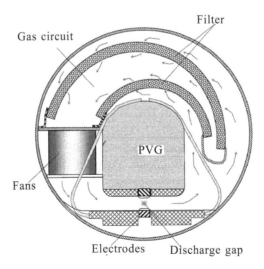

Fig. 1.7. Schematic of a compact P–P non-chain HF (DF) closed-loop laser [78] (PVG – pulsed voltage generator).

The high stability of the discharge is also ensured by the use of semiconductor electrodes [74, 27] or anisotropic-resistive cathodes, which provide high stabilization of the discharge [79, 80]. However, high heat losses in the material of the anisotropic-resistive cathode during long-term operation can lead to damage to the cathode, therefore, it is desirable to study in more detail the possibilities to increase the stability and uniformity of the OCP through the use of semiconductor and anisotropic-resistive cathodes.

To conclude this section, we note that success in the implementation of high-energy CLs implies the following task – the conversion of the radiation frequency of the HF(DF) laser to obtain lasing in a wider spectral range. The problem of finding effective frequency converters of high-energy CLs has been dealt with for a long time [3]. In [81], the second harmonic of DF laser radiation was generated in a $ZnGeP_2$ crystal; to obtain wavelengths with $\lambda < 2.5$ μm, there is also the possibility of generating overtone lasing [3]. A parametric light generator based on a CdSe crystal pumped by an HF laser was created in [82]. Recently, there has been great interest in the search for powerful sources of radiation from the terrahertz region of the spectrum [83]. To obtain lasing in this long-wavelength region of the IR spectrum, lasers based on HF rotational transitions were studied [84–86]. However, the most urgent for those application areas in which the HF(DF) laser was used is the search for the possibility of creating laser systems for generating powerful, tunable in the spectral range of 4–6 μm coherent radiation [87–91].

Thus, to increase the energy characteristics of non-chain HF(DF) lasers initiated by volume independent discharge (VID), it is necessary to search for new principles for the formation of a scalable volume discharge. Another important problem is the search for methods for producing high energy, pulsed, and medium power in the spectral region with $\lambda > 4.1$ μm, where the generation efficiency of electric-discharge non-chain HF(DF) lasers is low.

Therefore, we focused on solving two problems:

1. The formation of a scalable SSVD in strongly electronegative gases and the creation of wide-aperture high-energy non-chain HF (DF) lasers.

2. Expanding the generation spectrum of a non-chain HF(DF) laser and creating lasers with high energy per pulse, pulsed and average power in the spectral range $\lambda > 4.1$ μm.

2

Self-initiated volume discharge in working environments of non-chain HF(DF) lasers

In the first chapter, various types of non-chain HF(DF) lasers initiated by volume self-sustained discharge (SSVD) were considered and it was shown that to increase the output energy of these lasers, it is necessary to increase the volume occupied by the volume discharge plasma. The implementation of traditional approaches to the formation of SSVDs in SF_6-based gas mixtures in large discharge volumes faces major challenges. However, it was shown in [68, 69] that, in order to obtain SSVD in SF_6 and SF_6 mixtures with hydrogen donors (deuterium), there is no need for preliminary ionization of the gas if small-scale, ~50 μm, heterogeneities are applied to the cathode surface [92]. This form of SSVD was called self-initiated volume dischage (SIVD) [69]. This chapter presents the results of studies of SIVD in SF_6-based gas mixtures obtained in [68, 69, 92–97, 99–102].

2.1. SIVD in highly electronegative gases (methods of preparation and basic properties)

2.1.1. Description of experimental facilities and experimental techniques

Most of the experiments were carried out in a cylindrical dielectric discharge chamber filled with gas mixtures with a total pressure of $p = 1 \div 250$ mm Hg. SSVD was ignited in a discharge gap (DG) with different configuration of electrodes, shown schematically in Fig. 2.1,

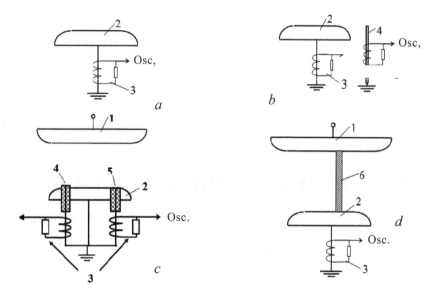

Figure 2.1. Disharge gap (DG) with different configuration of electrodes: 1 – anode; 2 – cathode; 3 – Rogowski belt; 4 – initiating electrode; 5 – control electrode; 6 – plexiglass plate.

with $pd = 0.02 \div 0.9$ cm · atm (p is the mixture pressure, d is the interelectrode distance). When studying the characteristics of SSVD in working mixtures of non-chain HF lasers, mixtures of technically pure SF_6 with hydrogen or hydrocarbons (C_2H_6, C_3H_8, c-C_6H_{12}) were injected into the chamber. In order to reveal some features of the development of SIVDs in working mixtures of non-chain HF laser, other gases were also introduced into the discharge chamber (CF_4, C_3F_8, c-C_4F_8, CCl_4, C_2HCl_3, CO_2, Ar, He, Ne, N_2). The electric circuits of the experimental installations are shown in Fig. 2.2.

The stability, electrical characteristics of the discharge as well as its dynamics were studied in the system of flat electrodes, schematically depicted in Fig. 2.1*A*. The discharge ignited at $d = 2.6$ cm in a DG with flat Al electrodes (Cu, Fe, Mg, Ti, Pb, SiC): diameter 6 cm and 12 cm, rounded around the perimeter with a radius of 1 cm, or the same flat electrodes 6 cm. The surface of the cathode, including rounding, was subjected to sandpaper or sandblasting to create small-scale (50 μm) inhomogeneities on the cathode surface, which ensures the production of a volume discharge in the form of SSVD in the non-chain HF(DF) laser mixtures. To study the spatial-temporal evolution of the SSVD, an artificial line with a variable number of cells (Fig. 2.2 *a*) was discharged onto this gap, which

allowed changing the voltage exposure time on the discharge gap within 30÷400 ns at a constant current and tracing the discharge development in time and space by photographing it at various voltage durations. In all other studies of SSVD using this and other systems of electrodes, a capacitor was discharged at the discharge gap (Fig. 2.2 *b* and Fig. 2.2 *c*). On the side the discharge gap in Fig. 2.1 *a*, was illuminated by a low-current (no more than 3 A) spark in a quartz shell. Such a spark could not provide volumetric photoionization of the gas, but due to the photoemission of electrons from the cathode, it allowed to stabilize the spread of the breakdown delay times of the discharge gap.

Photographing the discharge while varying parameters such as pumping capacity $C = 0.25÷15$ nF, inductance of the discharge

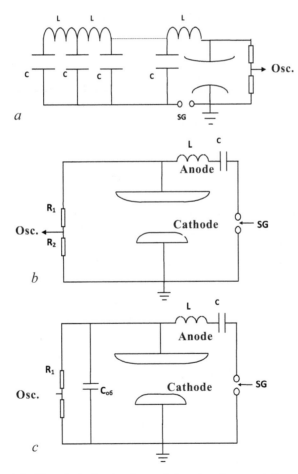

Fig. 2.2. Electrical circuits for studying the characteristics of SSVD.

circuit $L = 0.3 \div 16$ µH and charging voltage $U = 20 \div 50$ kV, allowed us to investigate the dependence of the surface density of a cathode spot (CS) on the energy input into the SSVD plasma, the duration and amplitude of the discharge current. During experiments on an oscilloscope with a bandwidth of 100 MHz (Tektronix TDS-220, 1012V), the current and voltage of the discharge were recorded using a calibrated shunt (or the Rogowski coil) and a high-voltage voltage divider. The energy W entered into the discharge plasma was determined as the result of numerical integration over the time of the product of current and voltage across the discharge gap (oscillograms of voltage and current were output to a computer), i.e. $W = \int_0^T U \cdot I dt$ where T is the duration of the current pulse, I and U are the discharge current and voltage at the discharge gap, respectively

The limits of stability of SSVD were controlled by the dependence of the maximum energy stored in the capacitor $CU^2/2$, at which the discharge does not end with a spark, on the circuit parameter, defined as the half-period of the short-circuit current of the discharge gap. The higher $CU^2/2$ at a fixed T, the higher the stability, and also the higher T at a fixed value of $CU^2/2$, the higher the stability [94–97]. Thus, in the stability studies, the parameter $CU^2/2$ was used to estimate W. The error of such a determination of W under the experimental conditions did not exceed 20%. The effect of the sharpening capacitance C_{sh}, which is connected in parallel with the discharge gap, on the stability of the SIVD in SF$_6$ was also studied. The electrical circuit of the installation for this experiment is shown in Fig. 2.2c. It should be noted that, in contrast to the values of C_{sh} used in real HF(DF) lasers, in these experiments the value of C_{sh} was approximately an order of magnitude smaller than the pumping capacitance. The choice of such a value of C_{sh} was due to the fact that the connection to the charging gap (through the lowest possible inductance) of such a capacity made it possible to obtain rather high peak values of the pump power (P_p). Since P_p (t) $= U_{pl}(t)*I(t)$, and the voltage across the plasma is equal to its quasi-stationary value $U_{pl} \approx U_{qs}$ (close to static breakdown), then in experiments with C_{sh} the stability of SSVD with respect to sharp perturbations of the discharge current was investigated.

The study of the dynamics of SSVD development was carried out in several electrode systems, including the use of a high-speed camera. Figures 2.1b and 2.1c show the experimental setup with a sectioned cathode. Two variants of the sectioned cathode were used.

The main cathode and anode were the same as in Fig. 2.1*a*, but in the first case (Fig. 2.1*b*) parallel to the main cathode at a distance of 5 mm horizontally from its edge a grounded conductor was connected (wire in polyethylene insulation with core diameter 1.5 mm). The vertical distance between the surface of the anode and the tip of the conductor (initiating electrode) was chosen so that the breakdown of the discharge gap occurred first on the conductor. In another embodiment of the sectioned cathode, the sections were conductors in polyethylene insulation with an inner core diameter of 1.5 mm, inserted into the holes in the main cathode, with the conductor 1 being the initiator, and its end face 1–3 mm protruded above the cathode surface, and the end face the control conductor 2 was located at the level of the surface of the main cathode (Fig. 2.1*b*). The initiation and control electrodes were located at the maximum possible distance from each other.

The setup diagram for studying the dynamics of SSVD formation using a high-speed camera is shown in Fig. 2.3. A SSVD with a total duration of ~500 ns ignited in an SF_6:C_2H_6 = 10:1 mixture at a pressure of 33 mm Hg and an interelectrode distance of 4 cm. The electrodes were a strip of foiled textolite 16 cm long (cathode mounted on the edge and a disk anode 6 cm in diameter, rounded around the perimeter with a radius of 1 cm. The breakdown was

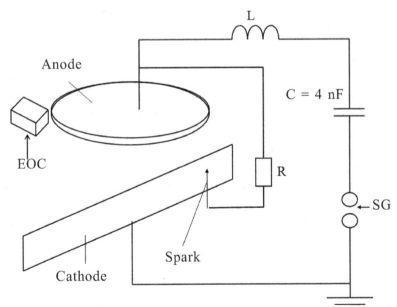

Fig. 2.3. Diagram of the experimental setup for studying the dynamics of the formation of SSVD [100].

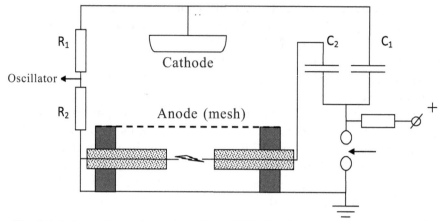

Fig. 2.4. Scheme for studying the effect of UV illumination on the characteristics of SSVD.

initiated at the edge of the discharge gap by a spark limited by the resistance R = 900 Ohm. The luminescence of the SSVD was recorded with an electron-optical camera (EOC) with an exposure time of 20 ns, triggered with a variable delay T relative to the start of the discharge [100].

The influence of the non-uniform distribution of the electric field in the DF on the characteristics of SIVD was studied in the electrode system, schematically shown in Fig. 2.1d. In the discharge gap in Fig. 2.1d, flat electrodes (the same as in Fig. 2.1a were bridged by a dielectric plate with a thickness of 2 mm and a transverse size of 5 cm.

Figure 2.4 shows the setup diagram, which investigated the effect of UV pre-ionization on the characteristics of SIVD (SSVD) in SF_6 [101]. The discharge was ignited in an SF_6:C_2H_6 = 10:1 mixture at a pressure of P = 33 mm Hg and the interelectrode distance d = 4 cm between the disk cathode with a diameter of 6 cm, rounded around the perimeter with a radius of 1 cm and subjected to sandblasting, and the grid anode, behind which there were placed 4 parallel spark gaps for the implementation of UV illumination of the discharge gap. The capacitor capacitances in the circuit in Fig. 2.4 were C_1 = 4–15 nF, C_2 = 680 pF (capacitor C_2 consisted of 4 capacitors of 170 pF, each of which was discharged to its own spark gap).

The uniformity of the discharge (the degree of overlap of diffuse channels, determined by the surface density of the CS) was estimated from photographs of the discharge gap. The stability of

SSVD in modes with UV illumination and without illumination was characterized by the dependence of the limiting energy $W_{lim} = C_1U^2/2$ stored in capacitors, at which SIVD still does not turn into a spark, on the parameter $T = \pi(LC_1)^{1/2}$, which characterizes the current duration discharge, where U is the voltage at C_1, L is the inductance of the circuit. The larger the W_{lim} value at a given T, the higher the stability.

2.1.2. General characteristics of SIVD

Photographing the discharge gap showed that the SIVD in SF_6 and its mixtures with different gases was a set of diffuse channels extending to the anode, tied to brighter and more dense channels that sprout from the cathode spot. A photo of the SIVD in an SF_6:C_2H_6 = 10: 1 mixture at a total pressure of 33 mm Hg, obtained in a system of flat disk electrodes (Fig. 2.1a) is shown in Fig. 2.5a. In appearance, SIVD does not differ from the usual SSVD with preionization. One should immediately dwell on the assumption that the corona with microinhomogeneities on the cathode surface plays a significant role in initiating the SSVD. If the corona effect is so significant that it allows one to obtain the necessary initial electron concentration n_e^{min} in the gap volume, then it should provide a stable (and very small) delay time of the electric breakdown of the discharge gap when a voltage pulse is applied to it. Estimation of the time of breakdown formation when applying a voltage pulse to the discharge gap with an overvoltage coefficient $K = 3$ ($t_f \sim \ln N_{cr}/\alpha v_e$, where $N_{cr} \approx 10^8$ is the number of electrons in the avalanche at which the avalanche-streamer transition occurs, α is the Townsend ionization factor, v_e – electron velocity [44]), gives the value $t_f \sim 10^{-8}$ s. However, it turned out that without the illumination of the discharge gap with a low-current spark, the delay in the breakdown at $K = 3$ can exceed 1 μs (both in pure SF_6 and with the addition of hydrocarbons). This is manifested even with an increase in the cathode area by an order of magnitude in the change in the amplitude of the breakdown voltage from pulse to pulse. Therefore, the corona from the cathode surface cannot provide the initial concentration of electrons necessary for obtaining SSVD from the volume of the dischage gap. As already mentioned, the photo SIVD (Fig. 2.5a) is no different from the SSVD with preionization. Waveforms of voltage and current of the SIVD are typical for SSVD waveforms in electronegative gases. In order to reduce the spread of the breakdown time of the discharge gap, the

 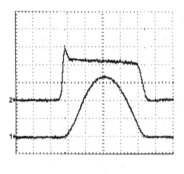

a)   b)

Fig. 2.5 (a) Photograph of an SIVD in a system of flat electrodes, SF_6:C_2H_6 = 10:1, P = 33 Torr; b) Typical oscillograms of current (1) and voltage (2) of the SIVD, sweep 100 ns/div.

cathode was irradiated with a local spark emitted from a quartz tube.

Figure 2.5*b* shows typical oscillograms of the current pulse and voltage across the discharge gap when the capacitor is discharged (corresponds to the conditions in which the picture of the volume discharge in Fig.2.5*a* is taken), from which it is clear that energy is introduced into the SIVD plasma at a quasi-stationary voltage U_{qs} (determined by at the maximum current).

Measurements of the value of U_{qs} showed that in mixtures of SF_6 with hydrocarbons with a content of the latter no more than 17%, it weakly depends on the partial pressure of the hydrogen carrier. The dependence of U_{qs} on $p \cdot d$ at $p \cdot d = 0.05 \div 2.5$ cm $\cdot$ atm and the specific energy contributions to the discharge plasma up to 0.2 J/cm^3 is well described by the expression

$$U_{qs} = A + B \cdot p_{SF_6} \cdot d \qquad (2.1)$$

where p_{SF_6} is the partial pressure of SF_6 in the mixture. The constants A and B for pure SF_6 and mixtures of SF_6:C_2H_6 = 10: 1, 10: 2 are, respectively: A = 0.72; 0.79; 1.1 kV, B = 92.7; 94.8; 96.4 kV/cm$\cdot$ atm. The values of B obtained in these experiments, within 8%, are close to the known [13] critical value of the reduced electric field strength $(E/p)_{cr}$ = 89 kV/cm $\cdot$ atm for SF_6, and it is this value that can be used to calculate the characteristics of the SSVD (SIVD) in working mixtures of non-chain HF(DF) lasers, since the content of hydrocarbons (carbon deuterides) usually does not exceed 10% [99]. Note also that for sufficiently large d, U_{qs} is close to the static breakdown voltage in SF_6 in a uniform electric field.

2.1.3. Effect of UV illumination on SSVD characteristics

The effect of UV preionization on the characteristics of SSVD in
SF_6 was studied on an experimental setup, which is schematically
shown in Fig. 2.4.

Figure 2.6 shows the dependences of W_{lim} on the parameter of
the circuit T, taken with and without UV illumination. As can be
seen from Fig. 2.6, the limiting energy is weakly dependent on
the illumination in the entire investigated range of changes in the
duration of the SIVD current. Typical oscillograms of the voltage
of a SIVD plasma under irradiation of the gap with UV radiation
and without using UV illumination from spark sources are shown in
Fig. 2.7. Figure 2.8 shows photographs of the glow of the cathode
discharge region, obtained at different values of T and the energy
W_{in} introduced into the plasma with the inclusion of UV illumination
and in its absence. Photographing was specially carried out so
that less diffuse channels than the discharge gap (linked to the
discharge gap) were not registered. The discharge gap surface density
characterizes the degree of overlap of the channels attached to them

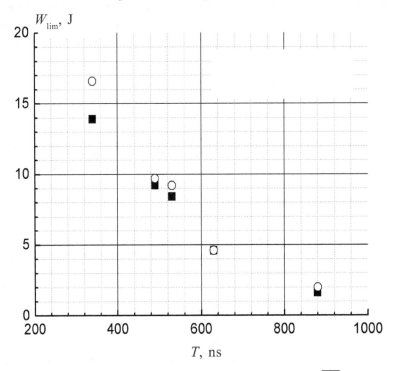

Fig. 2.6. Dependence of W_{lim} on the circuit parameter $T = \sqrt{LC}$: ■ – with UV
illumination; O – without UV illumination.

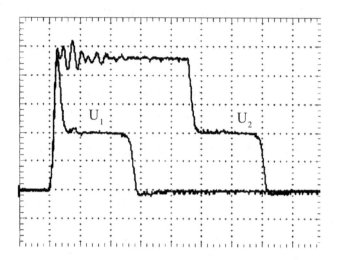

Fig.2.7. Typical oscillograms of voltage U_1, U_2 in the discharge gap (sweep 100 ns/div.) For two discharge pulses in the absence of UV gap illumination when the capacitor discharged at the discharge gap.

sufficiently long duration of the discharge current $T = 260 \div 270$ ns, the discharge gap density and the uniformity of their distribution over the cathode surface are almost independent of UV illumination. Both with illumination and in its absence, the total amount of discharge gaps and their density increases with the energy introduced into the plasma. In the case of a short discharge, $T = 130$ ns, (Fig. 2.8a, 2.8b) in the absence of illumination, the area in which the spots burn is only a part of the cathode surface [101].

Figure 2.9 shows the dependences of the total cathode spot number, N_s, on the energy W_{in} entered into the plasma at a fixed SSVD current duration ($T = 260 \div 270$ ns), taken with UV illumination and in its absence. From Fig. 2.9 one can see an increase in the number of cathode spots with an increase in the energy input, and the illumination has practically no effect on the value of N_s. The decrease in N_s at high energies is due to the onset of contraction of the SSVD as a result of the sprouting of the spark channel from the cathode spots and the discharge current to it. From the presented material it follows that the UV illumination does not have a significant effect on the stability and uniformity of the SSVD in the working mixtures of an electric-discharge non-chain HF laser.

Note once again (see Section 1.3) that in wide-aperture non-chain HF(DF) lasers, UV radiation in principle cannot provide volume

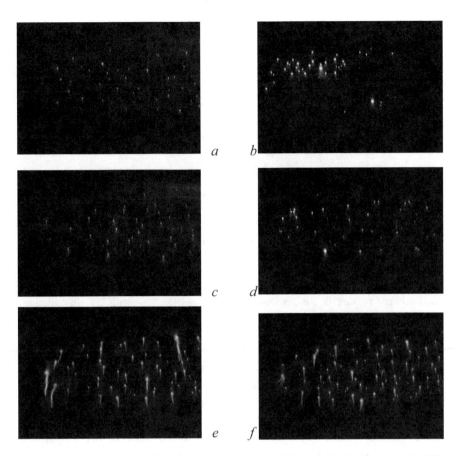

Fig. 2.8. Photos of the glow of the near-cathode region of the discharge with UV illumination (*a, c, e*) and without it (*b, d, f*): *a, b* – W_{in} = 1.4 J, T = 130 ns; *c, d* –W_{in} = 1.4 J, T = 270 ns; *d, f* –W_{in} = 10 J, T = 260 ns.

ionization of the working medium due to its strong absorption in SF_6, i.e. in this case, the term 'pre-ionization' in its usual sense [48] is not applicable at all. Since at the typical for non-chain HF laser partial pressure of SF_6 in a mixture of $P \approx 60$ mm Hg the photon path length with a quantum energy of ~11 eV close to the ionization potentials of 'heavy' hydrocarbons (hydrogen donors) does not exceed 5 mm (see Fig. 1.5). This is clearly not enough to talk about the possibility of 'preionizing' the medium with UV radiation even in compact HF lasers. However, when the photon energy <5 eV, the absorption of UV radiation in SF_6 is insignificant [56], and the UV illumination of the discharge gap even low-current, ~1 A, with a spark allows, due to the photoelectric effect on the cathode, to stabilize the electrical breakdown characteristics of the discharge gap. From Fig. 2.7 it

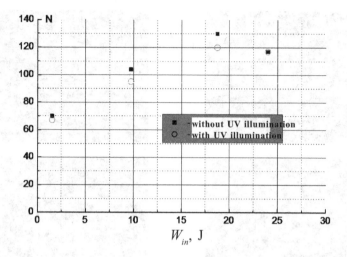

Fig. 2.9. Dependence of the number of cathode spots N_s on the energy introduced into the discharge plasma W_{in}: ■ – without UV illumination; ○ – with UV illumination.

can be seen that even with more than double overvoltage on the discharge gap (relative to the voltage in the quasi-stationary phase of the SSVD), the delay time of the electrical breakdown for two pulses in the absence of UV illumination reaches ~450 ns. As was shown in special studies [101], an increase in the duration of the rise time of a voltage pulse in the absence of UV illumination leads not only to a temporary spread, but also to a spread in the amplitude of the breakdown voltage of the discharge gap from pulse to pulse, and this, in turn, unlike electrical circuits with a discharge on the discharge gap capacitor or Arkadyev–Marx pulse voltage generator is accompanied by a large variation in the amplitude of the current and, accordingly, the energy input to the discharge plasma. A similar situation is observed in discharge gaps with a cathode area of $S \leq 300$ cm^2 and interelectrode distances of $d \leq 5$ cm, especially in lasers the electrodes of which have a special profile to ensure the uniformity of the electric field in the gap and a polished surface (we recall that in the experiments described here the surface of which was sandblasted to apply small-scale inhomogeneities on it). The instability of the current and voltage of the SSVD leads to non-reproducibility of the output HF laser energy from pulse to pulse, therefore, to stabilize the electrical breakdown characteristics of the discharge gap in such systems, it is usually sufficient to illuminate the cathode with a spark [20]. With an increase in the interelectrode distance and the cathode area, as well as with the use of electrodes with a large edge amplification of the electric field, the problem of stabilizing the characteristics of electrical breakdown disappears [94]. Thus, it follows from the above discussion that the role of UV illumination in non-chain HF(DF) lasers is reduced to stabilization of the delay time and voltage amplitude of the pulsed breakdown of the discharge gap due to the photo effect on the cathode. When the laser aperture is $d \geq 5$ cm, the photoionization of the medium by UV radiation is in principle not capable of providing the concentration of initial electrons in the discharge gap volume necessary for ignition of the SSVD due to the strong absorption of UV radiation in SF_6. As noted above, with a small duration of the discharge current, $T \leq 150$ ns, typical for installations with small volumes of the active medium (~100–200 cm³ [71, 74]), in the absence of UV illumination, in addition to the instability of the electrical breakdown of the discharge gap, the distribution of the discharge current density over the surface (especially along the length) of the cathode due to the final propagation velocity of the discharge in the direction transverse to the electric field after the local initial breakdown of the discharge gap. This naturally leads to a decrease in the output characteristics of the HF laser by reducing the length of the active medium and increasing the local energy release and leveling the current density distribution along the cathode when turning on the UV illumination (in this case it is advisable to distribute the sources of UV radiation along the cathode), followed by an increase in output energy, which is perceived as a result of 'preionization' of the environment.

2.1.4. Investigation of the stability of SIVD in SF_6 and mixtures based on it

The main part of the experiments to study the stability was performed for mixtures of 30 mm Hg. SF_6 + 3 mm Hg impurities in a consistent mode, when the magnitude of the voltage U applied to the dicharge gap (capacitor charging voltage) is $2 \cdot U_{qs}$ [95].

Figure 2.10a shows the dependences of the $CU^2/2$ parameter on SF_6 and SF_6 mixtures with H_2, N_2, Ar, C_2H_6 with an interelectrode distance of $d = 4$ cm. As can be seen from this figure, the stability of SIVD deteriorates noticeably when added to SF_6 Ar, H_2 and N_2, which [95, 103] were associated with the stepwise ionization processes of these gases due to the high average electron energy in SF_6. The addition of C_2H_6 (or any other hydrocarbon or carbon deuteride) is accompanied by a tremendous increase in stability; at $T < 300$ ns,

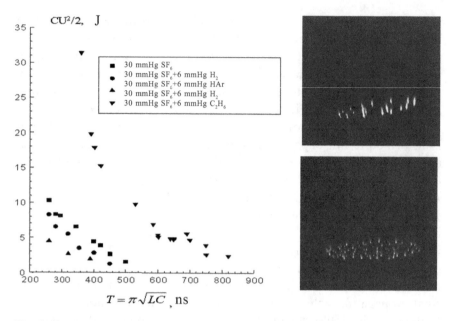

Fig. 2.10; a) Dependences of the limiting value $CU^2/2$ on the parameter T of the circuit; b) photo of the cathode emission in pure SF_6; c) photo of the cathode emission in the $SF_6:C_2H_6 = 10:1$ mixture. The SF_6 pressure was 30 mm Hg.

the energy input to the SIVD plasma exceeds 600 J/l. Figure 2.10 (*b, c*) shows photographs of the emission of the discharge gap in pure SF_6 and in an $SF_6:C_2H_6 = 10:1$ mixture with the same energy introduced into the SIVD plasma.

The photographs of the cathode glow show that in the SF_6: $C_2H_6 = 10:1$ mixture as compared to pure SF_6, the cathode spot density on the cathode increases by 5–6 times with a significant decrease in the length of the bright part of the channels, and the current and the SSVD voltage do not noticeably change (oscillograms are identical). Therefore, it was suggested in [95] that it is an increase in the surface density of a cathode spot, accompanied by a corresponding decrease in the current through the cathode spot, that is the reason for such a significant increase in the stability of the SIVD when hydrocarbons or carbondeuterides are added to SF_6. As will be shown in the following sections (see Chapter 4), this is not quite the case. The increase in stability is due to the change in the kinetics of plasmochemical processes occurring in the plasma of a single diffuse channel. Figure 2.11 shows the results of a study of the effect of the sharpening capacitance C_{sh}, which is connected in parallel with the discharge gap, on the stability of the SIVD in SF_6

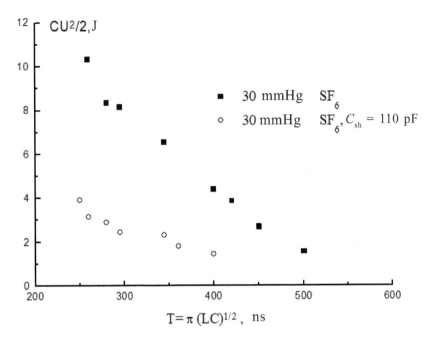

Fig. 2.11. Dependence of the limiting value $CU^2/2$ on the circuit parameter T.

(dependences of $CU^2/2$ on T obtained for $d = 4$ cm). As follows from the dependence of the $CU^2/2$ parameter in Fig. 2.11 on T, the connection $C_{sh} = 110$ pF strongly reduces the discharge stability, while this effect is absent in the SF_6 mixture with hydrocarbons. Thus, SIVD in the SF_6 mixtures with hydrocarbons (carbondeuterides) is less sensitive to sharp current fluctuations than in pure SF_6.

Of natural interest was the study of the influence of the state of the surface and the material of the cathode on the stability of SIVD. It turned out that SIVD in the gaps with cathodes having a polished surface is less stable than in the gaps whose cathodes were subjected to sandpaper or sandblasting (surface heterogeneity of 50 μm). Characteristically, the number of cathode spots in the SF_6 mixtures with hydrocarbons on a polished electrode is significantly less than on the cathode treated with sandpaper, while in pure SF_6 the number of cathode spots on the polished cathode surface treated with emery paper is approximately the same [20]. The discharge stability when using sandpaper-treated (or sandblasted) cathodes of Al, Cu, Fe, Ti and Mg is approximately the same (the experiments were carried out at $d = 4$ cm). Note also that in our experiments, no significant changes in the parameters of the stability of the SSVD were observed when replacing a cathode with a rounded edge with a cathode with a sharp edge [94, 20].

A significant increase in the stability of SIVD was observed when usig volume-resistive cathodes made of a SiC plate [101, 104] with a specific resistance of $\rho \sim 25 \div 50$ Ohm·cm and cathodes made of anisotropic-resistive material [79, 98]. Figure 2.12a shows for comparison the dependence of the limiting electric energy that can be introduced into a plasma without a SSVD switching into a spark, W_{lim}, on the duration of the discharge current T, taken at an interelectrode distance of $d = 4$ cm with SiC and Al cathodes having the same dimensions 85 mm), as well as a cathode of anisotropic-resistive material (AR – cathode) with dimensions of the flat part of 5×5 cm. From Figure 2.12a it can be seen that at $T = 400$ ns the limiting energy with a volume-resistive cathode is at least 1.5 times greater than with solid metal. The use of a SiC cathode made it possible to obtain homogeneous SIVD in pure SF_6 and SF_6 mixtures with H_2, while SIVD contracted on a metal cathode under these conditions. Figure 2.12b shows a photograph of a SIVD with an SiC cathode obtained at $T = 350$ ns and a specific input energy $W_{in} = 80$ J/l. The stability of SIVD in electrode systems with an anisotropic-resistive cathode from [98] was even higher (see Fig. 2.12a).

2.1.5. Dynamics of the formation of SIVD

As noted above, outwardly and according to the oscillograms of

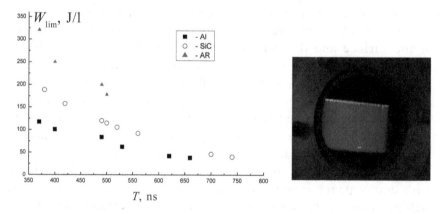

Fig. 2.12 a) Dependences of the limiting value of the energy introduced into the plasma W_{lim} without the transition of SIVD into a spark on T in a mixture of $SF_6:C_2H_6 = 30:3$ mmHg at $d = 4$ cm. b) SIVD photograph in pure SF_6 ($p = 30$ mmHg), $W_{lim} = 80$ J/l, $T = 350$ ns, SiC cathode.

voltage and current, SIVD is no different from the usual SSVD with preionization, ignited both in SF_6 and in other highly electronegative gases. The characteristic features of this form of a volume discharge are manifested in the dynamics of its development. Figure 2.13 presents the results of studies of the dynamics of the development of SIVD in the system of flat electrodes when discharging at the discharge gap of an artificial line with a variable number of cells: Fig. 2.13a – photographs of SIVD obtained at different current pulse durations (indicated under the frame); Fig. 2.13b – oscillograms of voltage (1) and current (2) of the SIVD when the line is discharged with the maximum number of cells in these experiments (20); Fig. 2.13c is the dependence of the total number of cathode spots on the cathode N_s on time t.

On a film with a sensitivity of 1600 ASA the first channels formed and then initiated the appearance of the following channels with a cathode spot, and the SIVD spread across the gap perpendicular to the direction of the electric field at a constant (equal to U_{qs}) voltage, gradually filling the entire discharge gap. The total number of spots on the cathode increases from the moment of the breakdown of the discharge gap almost in direct proportion to the time, i.e. taking into account the conditions of the experiment in proportion to the energy introduced into the discharge. Figure 2.13c shows the dependence of the number of cathode spot N_s on the duration of the voltage pulse t for the same current value. When the discharge duration is constant, the number of the cathode spots increases with increasing current amplitude.

As can be seen from Fig.2.13, in contrast to the SIVD with preionization, the SIVD initially ignites in the zone of maximum amplification of the electric field at the edge of the gap in the form of one or several diffuse channels attached to the cathode spots. The discharge glow in the rest of the gap at this point in time is not recorded even on a film with a sensitivity of 1600 ASA units. The first channels formed then initiate the appearance of the following channels with a cathode spot and the SIVD spreads across the gap perpendicular to the direction of the electric field at a constant (equal to U_{qs}) voltage, gradually filling the entire discharge gap. The total number of spots on the cathode increases from the moment of the breakdown of the discharge gap almost in direct proportion to the time, i.e. taking into account the conditions of the experiment in proportion to the energy introduced into the discharge. Figure 2.13c shows the dependence of the number of cathode spots N_s on

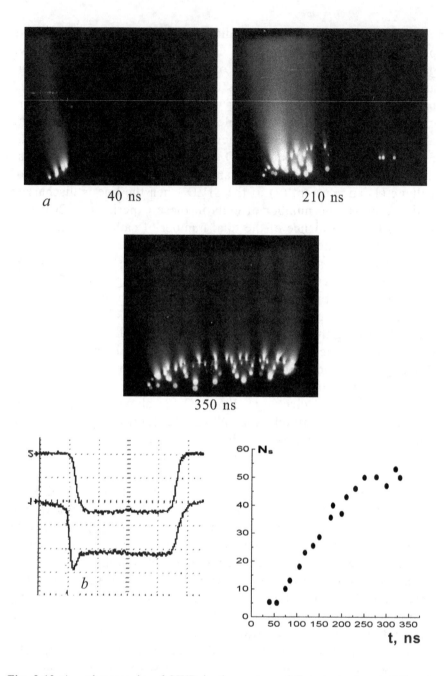

Fig. 2.13 a) – photographs of SIVD in the system of flat electrodes at different points in time, the spark illumination on the left; b) – oscillograms of voltage (1) and current (2) of SIVD when discharging over an artificial line gap, sweep 100 ns /div; c) – the dependence of the number of cathode spot, N_s on time t.

the duration of the voltage pulse t for the same current value. When the discharge duration is constant, the number of the cathode spots increases with increasing current amplitude.

Since the development of SIVD in the described conditions occurs at a constant current (with the exception of the leading and trailing edges of the pulse), with the advent of new channels, the current through the channels formed earlier should decrease, i.e. the extinction effect of a previously formed diffuse channel should be observed strange for independent discharges, despite the fact that during the entire process under consideration the voltage on the discharge gap exceeds the static breakdown voltage (taking into account the voltage drop on the cathode layer). The presence of this effect is indeed confirmed by experiments on the study of SIVD in a system of electrodes with a sectioned cathode (see Fig. 2.1c). Figure 2.14 shows oscillograms of currents through the initiating and control conductors (cathode sections). This figure shows that the current through the control conductor begins to flow with a noticeable delay relative to the current through the initiating conductor, corresponding to the propagation time of the SIVD over the cathode surface from the initiator to the control conductor, and by the time current appears through the control conductor, the current amplitude through the initiator decreases more than twice the magnitude of its magnitude. Thus, in the described experiment the effect of extinction of the diffusion channel formed by the first one is directly visible. The listed features of the development of SIVD indicate the presence of mechanisms for limiting the current density in the diffuse channel in SF_6 and mixtures based on it, which impede the flow of all energy through one channel. The existence of such mechanisms is also evidenced by the results of experiments in the system of electrodes with imitation of the appearance of the first diffuse channel by the initiating electrode located near the main cathode (Fig. 2.1b).

Figure 2.15a shows characteristic oscillograms of the discharge current through the initiating (1) and primary (2) cathodes, from which it can be seen that, indeed, the current through the initiating cathode appears much earlier than through the main and its maximum is reached at the time when the main current is still rising. Figure 2.15b shows the dependence of the fraction of energy Δ transmitted by the initiating cathode on the total energy W introduced into the discharge. The value of Δ monotonously decreases with increasing W, reaching saturation, which once again testifies in favour of the

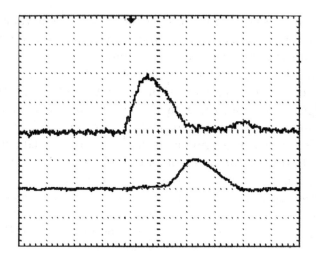

Fig. 2.14. Oscillograms of current through the initiating (upper beam) and control (lower beam) conductors (sweep 50 ns/div.).

existence of mechanisms that prevent the flow of all SIVD energy through one channel.

It is noteworthy that, starting from a certain specific value of the energy input to the SIVD plasma, a second maximum appeared on the oscillograms of the current through the initiating cathode.

Figure.2.16 shows the oscillograms of the voltage across the gap and the current through the initiating cathode, taken for a similar case. As follows from the oscillogram of the current in Figure 2.16, the conductivity of the first channel decreases from the amplitude value by almost an order of magnitude and only after some time begins to increase again. Thus, we have the effect of almost complete extinction of the first channel with the subsequent restoration of its conductivity (the effect of returning current to the channel). With a further increase in the energy input, the amplitude of the current in the second maximum begins to exceed the amplitude in the first, with a further increase in the energy input, the diffuse channel transforms into a spark. Note, however, that in this experiment, the imitation of the initial diffuse channel using a special electrode is not quite adequate to the actual conditions, since for a guaranteed start of discharge from the initiating cathode it was necessary to push it about 3 mm above the surface of the main cathode. Therefore, the simulated diffuse initial channel was noticeably distinguished by its

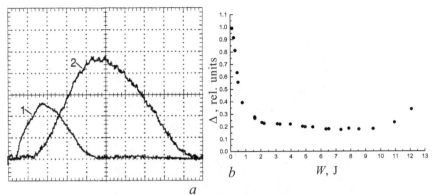

Fig. 2.15. *a*) – oscillograms of current through the initiating (1) and primary (2) cathodes, 39.7 A/div, sweep 50 ns/div; *b*) – dependence of the relative fraction of energy Δ transmitted through the initiating electrode on the total energy *W*.

brightness and transverse dimensions with respect to the channels in the main gap, while in real conditions this was not observed (Fig. 2.5*a*). It is obvious that in real conditions the value of Δ (Fig.2.15*b*) is significantly less than in the described model experiment.

The above-described dynamics of the development of SIVD is confirmed by studies carried out with the use of an electro-optical camera (EOC), which made it possible to obtain luminescence frames of SIVD plasma with an exposure time of 20 ns [100]. These experiments were carried out on an experimental setup, shown in Fig. 2.3. The electrode system was a system: a blade (cathode) – a plane (anode). The SIVD frames obtained with the help of the EOC when varying the camera start delay time relative to the start time of the discharge current are presented in Fig. 2.17.

The SIVD begins as a separate diffuse channel near the spark point. Then, next to the first channel, much less bright new channels appear, the number of which increases with time, and their brightness is gradually compared with the brightness of the first channel, with the most bright channels being closer to the primary channel. Over time, the luminous intensity of the first channel decreases noticeably, and all other channels become equally bright. With further development of SSVD, a gradual increase in the intensity of its luminescence is observed at the edge of the gap remote from the illumination point, however, at $T > 230$ ns, the luminescence again becomes uniform along the length of the cathode, and the luminescence in the initial channel is also restored. Further, the insulator of SIVD develops against the background of a general

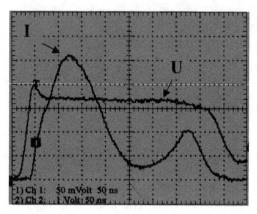

Fig. 2.16. Oscillograms of voltage across the discharge gap (*U*) and current through the initiating conductor (*I*), sweep 50 ns/div.

diffuse luminescence. The small black spot seen on all frames of Fig.2.17 is a defect of the camera.

2.1.6. Factors affecting the spatial homogeneity of SIVD in SF$_6$ mixtures with hydrocarbons

Obviously, the SIVD can only be conditionally attributed to the usual volume discharges, since this discharge is a set of diffuse channels growing from the cathode spot, i.e. it basically has a jet structure. Therefore, for the uniformity of this form of discharge, not so much the initial concentration of electrons, but the surface density of the cathode spot, which is largely determined by the conditions on the cathode surface, is more important. Since experiments with a sectioned cathode showed that the main discharge current flows through the cathode spot, it is necessary to know what determines the surface density of the cathode spot.

It should be noted that the interpretation of experiments with cathode spots is complicated by the fact that the state of the cathode surface changes during the experiment. With the increase in the number of the discharges, the number of cathode spots monotonously decreases, as can be seen from Table 2.1, where the dependence of the number of cathode spots (N_s) on the number of discharges *n* for Al and Cu cathodes is shown (the surface is treated with emery paper). Therefore, the comparison of various experimental data should be carried out for approximately the same values of *n*, or after the cathode training, when the dependence of N_s on *n* becomes weak. The data below is selected based on this requirement.

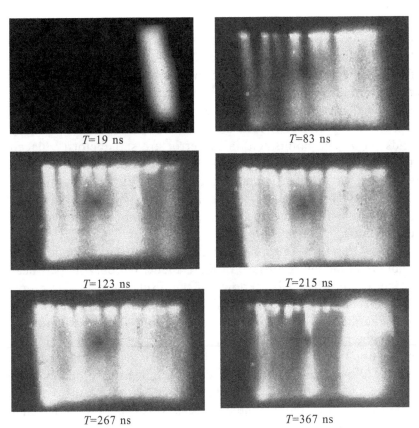

T=19 ns T=83 ns

T=123 ns T=215 ns

T=267 ns T=367 ns

Fig. 2.17. Photographs of SIVD with an exposure time of 20 ns, taken with a fixed time delay T relative to the moment of breakdown of the discharge gap.

Table 2.1. The dependence of the number of cathode spots on the number of discharges

Number of discharges n	0÷5	15÷30	100	1000	1100
Cu – number of cathode spots N_s	120	86	70	37	37
Al – number of cathode spots N_s	130	95	68	38	37
Pb – number of cathode spots N_s	61	53	28	–	–
SiC – number of cathode spots N_s	855	–	734	700	–

Figure 2.18 shows the dependence of N_s on the average density introduced into the plasma of energy W_{in} (measured in J/cm³), taken in the system of flat electrodes for different values of d and pressure p. The change in W_{in} during the experiments was carried out by changing the charging voltage and the magnitude of the charging capacitance, and the duration of the discharge current was also varied. From Fig.2.18 it is seen that for fixed p the value N_s weakly depends on d. The points within the error are satisfactorily placed on the linear dependence of N_s on W_{in}. Consequently, at a fixed pressure, the value of N_s is determined by the value of W_{in}, and not by the density and duration of the current. It is necessary, however, to note that the W_{in} value specifies the total number cathode spots on the cathode.

At low (less than 150 ns) duration of the current cathode spots may not cover the entire surface of the cathode, but only its part near the place of the primary breakdown of the discharge gap. Therefore, in short discharges, the local density of spots may be greater than with discharges with a longer current duration. The dependence of N_s on p is more complex. The value of N_s increases with increasing p and is not linear. In this case, there is also a decrease in the transverse dimensions of the cathode spot and diffuse channels at the anode, i.e. simultaneously with the growth of N_s, the volume occupied by each of the channels decreases. It can be assumed that this effect of an increase in p on N_s is associated with an increase in the electric field intensity (defined as $E_{qs} = U_{qs}/d$), at which the SIVD develops, and the value of which, as is known [44], significantly determines the probability of formation of a cathode spot. Given this assumption, the dependence of N_s on the parameter $W_{in} \cdot E_{qs}$ for different values of p and d is shown in Fig. 2.19. It is satisfactorily approximated by a linear function (almost directly proportional dependence)

$$N_s = A + B \cdot W_{in} \cdot E_{qs} \tag{2.2}$$

in which the constant B in turn is a function of the state of the cathode surface and the hydrocarbon content of the mixture.

Indeed, Figure 2.20 shows the dependences of the surface density of a cathode spot N_{sp} on the partial pressure of ethane ($P_{C_2H_6}$), obtained using cathodes with different surface states. An SSVD with a duration of 270 ns ignited in the discharge gap with the interelectrode distance $d = 4$ cm at a partial pressure of SF$_6$ of 30 mm Hg, the amplitude of the discharge current was ~1.7 kA). From these

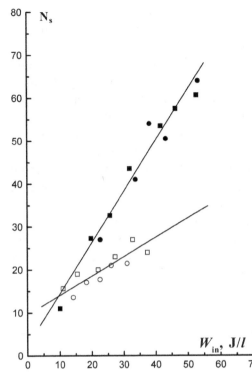

Fig. 2.18. Dependence of the number of cathode spots N_s on the specific energy input W_{in}. The mixture of $SF_6:C_2H_6 = 10:1$: ● – $d = 6$ cm, $P = 33.6$ mmHg; ○ – $d = 2$ cm, $P = 33.6$ mmHg; ■ – d = 6 cm, $P = 16.8$ mmHg. □ – $d = 2$ cm, $P = 16.8$ mmHg.

data it can be seen that the addition of ethane leads to an increase in the density of cathode spot and, as a consequence, an increase in the effective volume occupied by the discharge. Polishing and subsequent training of the cathode with approximately 100 discharges leads to a decrease in the number of gearboxes (curve 2 in Fig. 2.20). The stability of the SIVD is also reduced (see section 2.1.4 of this dissertation), but not as strong as the cathode spot density. It is noteworthy that an increase in the cathode spot density both in pure SF_6 and in an $SF_6:C_2H_6 = 10:1$ mixture resulted in the addition (up to 10% in pressure) of such a strongly electronegative gas, like C_2HCl_3. Apparently, in this case, an increase in the cathode spot density was due to an increase in the field strength at the cathode, since the addition of C_2HCl_3, in contrast to the addition of hydrocarbons, led to a noticeable increase in the discharge burning voltage.

An increase in cathode spot density when hydrocarbons are added to SF_6 is probably due to the low ionization potential of the latter, since a similar pattern is observed when SF_6 and other easily ionizable substances are added (C_2H_5OH, C_3H_7OH, xylene, tri-ethylamine, etc.). It should be noted that these compounds, when

adsorbed on the cathode surface, reduce the electron work function
of the metal (φ). For example, in the case of ethylene (Cu cathode),
the decrease in the electron work function of the metal $\Delta\varphi$ is 1.2
eV [105]. It is known [106] that from the areas of the cathode with
a reduced work function the current of field emission significantly
increases, which in turn can provoke the formation of a cathode spot
[105]. Let us show how the saturation character of the dependences
of the cathode spot density on the partial pressure of ethane can
be explained by the basis of the theory [105] – see Fig. 2.19 b.
Assuming that the CS will be initiated by an ecton arising from
the explosion of a local centre with increased electron emission,
it is possible to establish a functional relationship between the gas
pressure above the cathode surface and N_s. Here it is assumed that
it is ethane adsorption that lowers the work function of the electron,
causing an increase in the field emission current. Then the number
of cathode spots on a single surface will be proportional (for a given

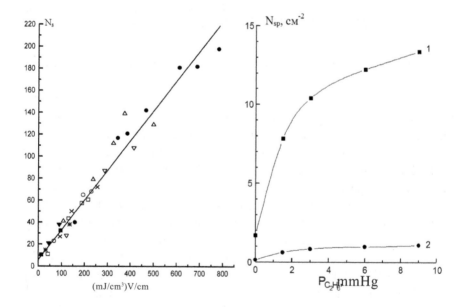

Fig. 2.19. a) – Dependence of the number of cathode spots N_s on the parameter
$W_{in} \cdot E_{qs}$. A mixture of $SF_6:C_2H_6 = 10:1$: ■ – $d = 6$ cm, $P = 33.6$ mmHg; – $d = 6$
cm, $P = 16.8$ mm Hg; ● – $d = 4$ cm, P = 23.3 mm Hg; O - $d = 3$ cm, P = 23.3
mm Hg.; O – d = 3 cm, P = 33.6 mm Hg; Δ – $d = 3$ cm, P = 50.4 mm Hg; ▽ –
$d = 3$ cm, $P = 67.2$ mm Hg; × – $d = 2$ cm, $P = 33.6$ mm Hg.

energy input) to the number of autoemission centers N_z – potential centres of ecton production, $N_{sp} \approx b \cdot N_z$. The number N_z of such centres per unit surface of the cathode will obviously depend on the state of the surface of the cathode and is determined by the dynamic equilibrium established in the processes of ethane adsorption on the center and its desorption from this site. (Sandblasting of the cathode surface obviously increases the number of sites with increased autoelectronic emission of electrons, the value of the coefficient b in this case will be higher). The rate of ethane adsorption on the cathode centers V_{ad} (i.e., the number of molecules adsorbed per unit of time) is proportional to the pressure of ethane and the number of free centers on the cathode surface. If the total number of centres is N_e, and when N_z centres are occupied in adsorption, then the number of centres remaining free is $(N_e - N_z)$. Therefore, $V_{ad} = k_{ad} P_{C_2H_6} (N_e - N_z)$. Adsorption is dynamically balanced by the desorption process. The rate of desorption is proportional to the number of adsorbed molecules: $V_{des} = k_{des} \cdot N_z$. At equilibrium, $V_{ad} = V_{des}$, therefore, $k_{ad} P_{C_2H_6} \cdot (N_e - N_z) = k_{des} \cdot N_z$. Let $k_{ad}/k_{des} = K_a$, we get:

$$N_{sp} \approx N_{sp}^0 + b \cdot N_{sp}^0 = + b \cdot N_e \cdot K_a P_{C_2H_6}/(1 + P_{C_2H_6} K_a) \qquad (2.3)$$

Here, N_{sp}^0 is the number of cathode spot in pure SF_6; b is a constant depending on the state of the surface and the cathode material, and the constant K_a also depends on the gas that is adsorbed on the surface of the cathode. In fact, expression (2.3) describes the adsorption isotherm for a monolayer (the Langmuir adsorption equation). Experimental dependences (points) shown in Fig. 2.20 with very good accuracy are approximated by the expression (2.3).

For practical applications, it is important to search for the optimal microstructure of the cathode surface, which could ensure maximum homogeneity of the SIVD. It should be borne in mind that the metal surface of the cathode (brass, stainless steel, duralumin, magnesium, titanium, lead were investigated) changes in the process of research, because the cathode spot is the result of an explosion of surface areas with increased emissivity (appearance of an ecton). Figure 2.20 shows photographs of the cathode surface area after these electrodes have worked in the discharge chamber. The processing of the electrode was as follows: in the SF_6:C_2H_6 = 30:3 Torr mixture. SIVD was ignited at this cathode with an interelectrode distance of d = 4 cm (flat diameter of the cathode 4.5 cm), the discharge duration was 270 ns, one pulse was 5 J, the electrodes were exposed to 150 such pulses.

Figure 2.20 shows the traces of craters that have formed after the formation of a cathode spot: the sizes of craters on the Al cathode were: diameter D_c = 25–50 μm, and depth h_c = 20–40 μm. On a polished cathode made of stainless steel, the scatter of the parameters of erosion defects (craters), formed in the process of passing the discharges, was much greater, D_c = 4–40 μm. Apparently, craters with a large diameter formed as a result of several discharges, localized in one place. This hypothesis is confirmed by the analysis of the morphology of defects with large sizes – they consist of a large number of small craters with sizes <10 μm. Understanding the complexity of the processes occurring on the surface of the cathode associated with the formation of craters: – the influence of the characteristics of the discharge pulse, the composition and pressure of the gaseous medium, and factors that are difficult to take into account, for example, the formation of dielectric films on the cathode during discharge in polyatomic gases (SF_6, C_2H_6, C_3H_8 etc.). We confined ourselves to the purely utilitarian question of how to treat the surface of the cathode so that the SIVD is uniform and retains its properties after passing through 10^3 or more discharges. For Al cathodes, it turned out that when applied to surface irregularities with sizes of ~50 μm, the cathode spot density on the surface after 100–200 discharges reaches its quasistationary value and, subsequently, does not change, since the defects resulting from the explosion-emission processes have the same spatial dimensions. If defects with a size of less than 30 μm were applied to the surface (emery paper with grade P40–P180 was used in treating the surface of the cathode), then, at first, the density of cathode spot was higher (the first 50 pulses), but after 200–300 shots it was already significantly lower than when the surface of the cathode was treated with emery paper labeled P22–P36. When the structure of defects on the cathode surface was too small (<10 μm), then after training the cathode by several discharges, the density of the cathode spots on this cathode was low, approximately the same as on the cathode subjected to optical polishing.

Attention should be paid to the high density of cathode spots on the cathode surface made of SiC (see Table 2.1). Two SiC cathodes were investigated — one was polished (the surface defect size did not exceed 10 μm), and the other had a rather rough surface with irregularities 30–50 μm in size, which made it possible to produce a

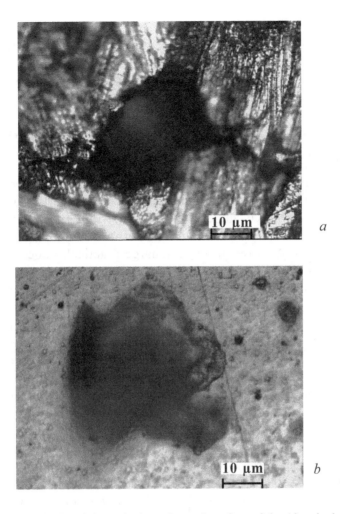

Fig. 2.20. Photographs of the cathode surface: *a*) surface of the Al cathode treated with sandpaper (grade P22); *b*) polished stainless steel cathode surface.

noticeably more uniform SIVD than on a polished cathode. (Table 2.1 shows the results of studies of the latter). The polished SiC cathode nevertheless made it possible to obtain a more stable and uniform discharge than on a metal cathode. The SiC cathodes showed high stability characteristics (density of the cathode spot on the cathode, the stability of SIVD) than metal. The lead cathodes proved to be the worst of all the studied electrodes – when using this material in the cathode, the stability and uniformity of the SIVD was the least, which was due to the absence of areas with high electric field amplification and high emissivity on the surface of such a cathode. Indeed, due to the low melting temperature of lead, microirregularities created using sandpaper and the edges of craters formed after the occurrence of a cathode spot looked rounded under a microscope, therefore the local amplification of the electric field near these defects was significantly lower than on steel or duralumin cathodes or duralumin.

2.1.7. Influence of the inhomogeneity of the electric field in the gap on the stability of SIVD

Due to the existence of mechanisms for limiting the current density in the diffuse channel, the stability of the discharge form under consideration weakly depends on the distribution of the external electric field in the gap, which is determined by the geometry of the electrodes. As a result, the discharge practically does not notice sufficiently gross defects of the cathode surface, such as small potholes, dents, and holes. Note that the electrodes in our experiments did not have any special profile that ensures the uniformity of the field; the edge of the cathode was rounded with a radius of $r << d$. Therefore, at the edge of the discharge gap, the electric field is always amplified; in the air in this system, the discharge was forced to the edge of the cathode and turned into a spark. However, the region with a high edge amplification of the electric field in SF_6 mixtures with hydrocarbons practically does not stand out against the general background of the discharge gap emission (Fig. 2.21). An experiment in the system of flat electrodes closed by a dielectric plate turned out to be very indicative in this respect (Fig. 2.1d). Figure 2.21 shows photographs of the SIVD in this system of electrodes obtained in air (Fig. 2.22 a) and in an SF_6:C_2H_6 = 10: 1 mixture (Fig. 2.21 b). As one would expect, the discharge in air develops as a spark breakdown of the gap over the surface of the dielectric plate, the volume phase is completely absent. In the SF_6: C_2H_6 mixture, the discharge has a pronounced volumetric character; in appearance and oscillograms, it is not at all different from the SIVD obtained in the same system in the absence of a plate. Thus, SIVD, even if it starts along the surface of the dielectric, thanks to the mechanisms of limiting the current density in the diffuse channel, is then forced into the depth of the discharge gap.

It is characteristic that the stability of SIVD does not depend on where the field is strengthened: at the edge of the cathode, at the edge of the anode, or at the edges of both electrodes, i.e. $h_c < h_a$, $h_c > h_a$, $h_c = h_a$, where h_c, h_a are respectively the transverse dimensions of the cathode and the anode, the same energy can be introduced into the SIVD plasma. Figure 2.22 presents the dependences of the maximum specific energy input on the SIVD current duration T, taken at an interelectrode distance of $d = 4$ cm and at a pressure of an $SF_6:C_2H_6 = 10:1$ mixture of 33 mmHg in the discharge gap with $h_c > h_a$, $h_c < h_a$, $h_c = h_a$ with a radius of curvature of the electrodes around the perimeter of $r = 1$ cm. The configuration of the discharge gap is also schematically shown in Fig. 2.22.

As can be seen from Fig. 2.22, the value really does not depend on the discharge gap configuration, which allows non-chain HF(DF) lasers to use, as an anode and cathode, flat electrodes rounded around the perimeter with a rather small radius $r \ll d$, h, i.e. apply extremely compact electrodes. This makes it possible to significantly reduce the dimensions of the discharge chamber of a non-chain laser and simplifies the problem of scaling its output characteristics.

2.2. Numerical simulation of the SIVD in SF_6-based gas mixtures

Carrying out full-scale studies of non-chain electric-discharge HF(DF) laser, whose working environment is a mixture of SF_6 with various hydrogen donors (deuterium), is a difficult task that requires significant material costs for its solution. These costs can be significantly reduced if at least partially direct experimental modelling is replaced by computer calculations. When calculating the characteristics of SSVD (SIVD) in SF_6, dozens of elementary processes should be taken into account, including an increase in electron losses due to their attachment to vibrationally excited SF_6

 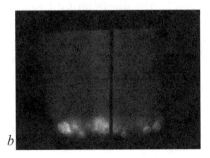

Fig. 2.21. Photographs of the discharge in a system of flat electrodes closed by a glass plate in air (*a*) and an $SF_6:C_2H_6 = 10:1$ mixture (*b*).

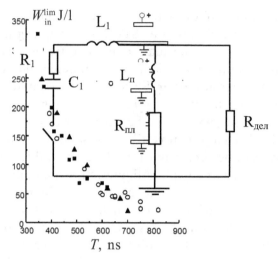

Fig. 2.22. Dependences of the upper limit of the specific energy input to the SIVD plasma on the circuit parameter $T = \pi(LC)^{1/2}$ in discharge gap with different configurations.

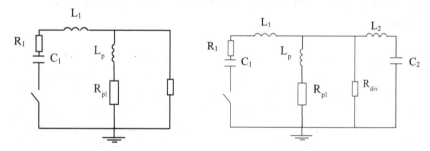

2.23. Equivalent electrical circuits for which numerical simulation was carried out: *a*) calculation of the discharge of capacitance; *b*) a capacitance is discharged on the DG, and a sharpening capacitance is connected parallel to the gap.

molecules, SF_6 dissociation by electron impact, recombination [98], stepwise SF_6 ionization processes and impurities, as well as the parameters of the external circuit. Accurate data on the rate constants of many elementary processes are not available, which does not allow us to single out the main processes that determine the characteristics of the SIVD and makes it difficult to verify the calculation results. On the other hand, to solve many problems it is often not necessary to determine the exact composition of the gas-discharge plasma – it suffices to know the conductivity of the discharge gap and the dynamics of the formation of the concentration of one or two components [99].

2.2.1. Calculation of the characteristics of SIVD in the working mixtures of HF laser

Numerical simulation of the characteristics of the SIVD in the working mixtures of the HF laser was carried out mainly to analyze the operation of various electrical pumping circuits used in laser installations created in the laboratory (determination of the discharge current, energy deposition in the SIVD plasma, etc.). The corresponding equivalent electrical circuits are shown in Figs. 2.24 and 2.25. The gas gap in the calculation was considered as a nonlinear resistance, the value of which was determined from the balance equations for electrons and ions. When comparing the experimental and calculated oscillograms, it was taken into account that real voltage dividers register the voltage not directly on the plasma, but through a certain inductance of the supply lines L_s. A detailed description of the method for calculating the characteristics of SIVD in SF_6 and its mixtures with hydrocarbons was presented in [99, 107]. We describe here only the key points of this technique. For definiteness, we will consider gas mixtures based on SF_6; however, this approach can also be used to calculate SSVD characteristics in a wide variety of mixtures based on highly electronegative gases (C_3F_8, C_2H_5I, C_3H_7I, etc.).

As mentioned earlier, for SIVD (SSVD) in SF_6, C_3F_8, C_3H_7I and their mixtures with hydrocarbons (H_2, D_2), there is a pronounced quasistationary phase of the discharge, followed by a rather sharp drop in voltage. The magnitude of the combustion voltage (voltage in the quasi-stationary phase) U_{qs} linearly depends on $p \cdot d$ [99] and is mainly determined by the partial pressure of the strongly electronegative component. For mixtures based on SF_6, the value of the reduced field strength (defining the slope of the straight line $U_{qs} = f(pd)$) is close to the critical value of the reduced field strength $(E/p)_{cr} = 89$ kV·cm^{-1}·atm^{-1} for pure SF_6 (value E/p, at which the coefficients of impact ionization α and dissociative attachment of electrons η equalize). This circumstance makes it possible to take into account when describing the basic characteristics of SSVD (voltage on plasma and discharge current) in SF_6 mixtures with hydrocarbons (hydrogen) at standard ratios of components for non-chain HF (DF) lasers, only SF_6 ionization and dissociative electron attachment to SF_6 molecules. The next important step, which allows us to significantly simplify the simulation, is that the dependences of the ionization and sticking coefficients, or more precisely the α/p and

η/p values (since they are functions of E/p (E/N), are approximated by linear functions).

$$\alpha/p = A_1 \cdot E/p + B_1, \qquad (2.4)$$
$$\eta/p = A_2 \cdot E/p + B_2. \qquad (2.5)$$

The possibility of using such an approximation instead of real dependences in calculating the SSVD (SIVD) characteristics is based on the fact that in strongly electronegative gases and mixtures based on them, the bulk of the energy stored in the pump source is introduced into the plasma at a voltage close to U_{qs}.

Thus, the general approach is as follows:

1) the discharge gap is considered as a resistive element.

2) For the mixtures used, the dependences $U_{qs} = (E/p)_{cr} \cdot p \cdot d + U_k$ are found experimentally, from which $(E/p)_{cr}$ is determined (U_k is the voltage drop on the cathode layer).

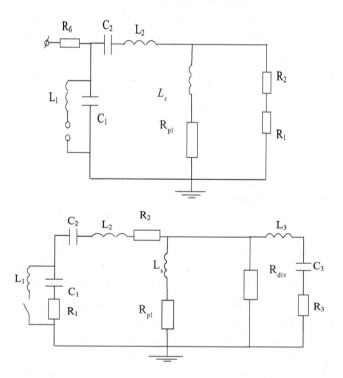

Fig. 2.24. Equivalent electrical circuits for which numerical simulation was performed: *a*) calculation of the discharge of the Fitch generator; *b*) a Fitch pulse voltage generator is discharged at the discharge gap, and a peaking capacitor is connected in parallel with the discharge gap.

3) In the region of the critical point, linear dependences are obtained for the reduced ionization and sticking coefficients; dependences of the form (2.4) and (2.5). For the mobility (velocity) of electrons, approximations are also prepared, with which, at E/p close to $(E/p)_{cr}$, it is possible to determine the velocity of the electrons with good accuracy.

4) The balance equations of electrons and ions are solved together with the Kirchhoff equations for an electrical circuit. The discharge current is determined by the formula $I = n_e \cdot e \cdot u_e \cdot S$, if the ionic component of the current can be neglected, and $I = (n_e \cdot u_e + n_+ \cdot u_+ + n_- \cdot u_-) \cdot e \cdot S$, when one needs to take this component.

Let us show how this technique is applied to the analysis of a simple circuit depicted in Fig.2.23a, where the capacitor is discharged into the gas gap (SF_6 is used as the working gas).

Since the ionic component of the current is small, we neglect it. Thus, the system of equations is solved:

$$dn_e/dt = \alpha_{eff} \cdot u_e \cdot n_e \qquad (2.6)$$
$$U_c + L \cdot dI/dt + I \cdot (R_d + R) = 0; \qquad (2.7)$$
$$dU_c/dt = -I/C; \qquad (2.8)$$
$$R_d = (U_c - L \cdot dI/dt - I \cdot R)/I; \qquad (2.9)$$
$$I = ne \cdot e \cdot u_e \cdot S - \text{discharge current}; \qquad (2.10)$$

In the calculation, the approximations from [13] are used:

$$\alpha_{eff} = A \cdot p \cdot (E/p - B) \qquad (2.11)$$
$$u_e = 1.6 \cdot 10^4 (E/p)^{0.6}, \qquad (2.12)$$

where $A = 0.027$ V^{-1}, $B = 89$ V·m^{-1}·Pa^{-1}; α_{eff}, u_e and E/p are measured, respectively, in m^{-1}, m/s and V · m^{-1}·Pa^{-1}.

Here, $\alpha_{eff} = \alpha - \eta$, e is the electron charge, n_e is the electron concentration, u_e is the electron drift velocity, C is the capacitor capacitance, L is the loop inductance, R is the resistance characterizing the losses in the circuit, R_d is the discharge resistance, U_c is the voltage on the capacitor, I is discharge current, p is the pressure of SF_6, S is the cathode area. The inductance L and the active resistance of the circuit R were determined experimentally by current fluctuations in the short circuit mode. The initial electron concentration n_0 was taken at the background level. The initial voltage on the capacitor was taken to be equal to the charging voltage, the initial current in the circuit was taken to be zero.

Figure 2.25 presents the calculated and experimental oscillograms of current and voltage in the discharge gap (p = 30 Torr, d = 5 cm) when simulating the circuit shown in Fig. 2.23a, while taking into account that the real divider registers the voltage on the discharge plasma plus a drop voltage on the inductance of the leads to the electrodes.

Thus, consideration of only the processes of ionization and sticking without taking into account such complex processes as step ionization, etc., makes it possible to obtain a real I–V discharge with good accuracy. The calculation gives good agreement with the experiment (the error is no more than 10%) with specific energy inputs to the SSVD plasma up to 150 J/l.

Let us now consider the limits of applicability of the numerical model described above and the main sources of errors arising from the simplifications used in its compilation. First, we note that these calculations do not take into account the magnitude of the cathode fall, which is usually small compared with $(E/p)_{cr} \cdot p \cdot d$, however, when analyzing small-aperture systems (d < 2 cm), this can lead to a noticeable discrepancy between experimental and calculated oscillograms. Secondly, in a gas-discharge plasma of highly electrically negative gases, the ion concentration is almost two orders of magnitude higher than the electron concentration, as a result, the ion component of the current makes a significant contribution to the total discharge current (especially at low pressures, when the recombination rate is low). Accounting for the ion component becomes simply necessary when analyzing circuits operating in an inconsistent mode, when a large residual voltage remains (close to U_{qs}). Thirdly, at large specific energy inputs (current densities), a significant dissociation of molecules occurs, which changes the electrical properties of the working mixture. Fourth, the simple model considered here, as well as any other more complex 0-dimensional model, does not take into account the change in the discharge volume in the process of energy input into the plasma, so it, in principle, cannot give a real description of the processes in the SSVD plasma (this is not required for the description of plasma electrical properties).

With the exception of the last fourth remark, the others can be easily taken into account in the model considered above only by a simple modification of the initial equations and by taking into account additional processes. However, this requires additional information on the values of the rate constants of the processes of

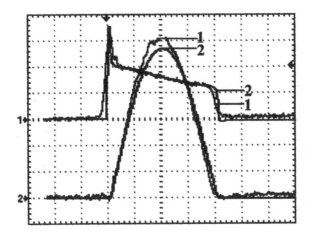

Fig. 2.25. Waveforms of voltage (upper beam) and current (lower beam); 1 – experiment; 2 – calculation; voltage 10 kV/div, current 200 A/div, sweep 50 ns/div.

dissociation, electron detachment from negative ions, ion–ion [98] and ion–electron recombination [100], etc. In addition, for specific mixtures it is required to verify the selected set constants. For example, for SF_6, the equations and constants will be as follows:

$$d_{ne}/dt = (\alpha-\eta) \cdot u_e \cdot n_e + K_d \cdot n_e \cdot n - \beta_{ei} \cdot n_e \cdot n_+; \quad (2.13)$$
$$dn_+/dt = \alpha \cdot u_e \cdot n_e - \beta_{ii} \cdot n_+ \cdot n - \beta_{ei} \cdot n_e \cdot n_+; \quad (2.14)$$
$$dn_-/dt = \eta \cdot u_e \cdot n_e - K_d \cdot n_e \cdot n_- \quad \beta_{ei} - n_+ \cdot n_-; \quad (2.15)$$

where $R_d = U_d/I$; – resistance of the discharge plasma; $I = (n_e \cdot u_e + n_+ \cdot u_+ + n_- \cdot u_-) \cdot e \cdot S$ is the discharge current; $E = U_d/d$ is the electric field in the gap; $dw/dt = I \cdot U_d/V_d$; $dp/dt = (1/qF) \cdot dw/dt$; $\beta_{ei} = 10^{-7}$ cm$^3 \cdot$ s^{-1}; $\beta_{ii} = 2 \cdot 10^{-8}$ cm$^3 \cdot$ s^{-1} are the rate constants of electron–ion and ion–ion recombination, respectively; $V_d = d \times S$ is the volume occupied by the plasma; w is the specific energy input into the SSVD plasma; $qF = 4.5$ eV is the price of the formation of a fluorine atom; $K_d = 10^{-7}$ cm^3 is the rate constant of the electron detachment reaction from a negative ion by electron impact.

The modified model was applied by us to calculate the discharge parameters in various laser facilities [99, 108]. In all cases, it gave a coincidence of the calculated oscillograms with the experimental ones within 8%, which is significantly better than it can be achieved in more complex models [109–111].

2.2.2. Modelling of the channel structure of SIVD in SF_6 and mixtures based on it

The dynamics of the formation of the jet (channel) structure of the discharge was modelled by a set of parallel-connected resistive elements corresponding to each channel. The resistance of the diffuse channel was defined as $R_c = U/(S \cdot e \cdot (n_i^- \cdot v_i^- + n_i^+ \cdot v_i^+ + n_e \cdot v_e))$, where U is the voltage across the discharge gap, S is the cross-sectional area of the channel, e is the charge of the electrons, n_i^-, n_i^+, n_e, v_i^-, v_i^+, and v_e are the concentrations and drift velocities of negative, positive ions and electrons, respectively. It was assumed that the sum of the areas of cross sections of all channels is equal to the cathode area. The particle concentrations were determined from the solution of the continuity equations for each particle type together with the Kirchhoff equations for the discharge circuit, similarly to [100, 112]. The inhomogeneity of the initial conditions for the development of channels along the length of the cathode was simulated by setting different values of the initial concentration of electrons in each of the channels. In addition to the processes of ionization of SF_6 by electron impact and the attachment of electrons to SF_6 molecules, the following processes were taken into account in the calculation [100].

1. *Dissociation of SF_6 by electron impact.* The number of dissociated molecules was determined as $N_d = W/q_F$, where W is the energy introduced into the discharge, and $q_F = 4.5$ eV is the energy cost of formation of the F atom from [51].

2. *Electron detachment from negative ions by electron impact.* It was assumed that negative SF_6 ions dominate in the plasma, since charge exchange reactions do not have time to occur during the discharge, and the cross sections for the formation of other negative ions by electron impact are too small [113]. The electron impact destruction rate constant of negative ions was estimated as $k_d = 10^{-7}$ cm$^3 \cdot$ s^{-1} under the assumption that its value should not be less than the value of the elastic scattering rate constant of the electrons on the SF_6 molecules [100].

3. *Dissociative electron–ion recombination.* The rate constant of this process, $\beta_{ei} = 10^{-7}$ cm$^3 \cdot$s^{-1}, was estimated under the assumption that positive SF_5^+ and $\beta_{ei} \sim T_e^{-1/2}$ ions dominate in the discharge plasma in SF_6, where T_e is the electron temperature [100].

4. Ion–ion recombination, the rate constant of which $\beta_{ii} = 2 \cdot 10^{-8}$ cm$^3 \cdot$ s^{-1} for close to critical values of E/N was taken from [98].

Figures 2.26*a* and 2.26*b* show calculated oscillograms of voltage U, total discharge current I, and current through a separate channel (the first for which the maximum value of the initial electron concentration was set) I_1 for two values of the energy density introduced into the discharge plasma: 80 J/*l* and 40 J/*l*. As can be seen from Fig. 2.26*a*, with a large input energy, the current through a separate channel has two maxima, which qualitatively agrees with the oscillogram of the current through the initiating conductor (Fig. 2.14) in the flat geometry of the gap and with the observed in-experiment redistribution of the intensity of the emission of channels when SIVD develops in the blade–plane gap (Fig. 2.17). With a decrease in the energy introduced into the plasma, the effect of current return disappears (Fig.2.26 *b*). Figure 2.27 shows the results of the calculation of the dynamics of current change in 4 diffuse channels. The inhomogeneity of the initial conditions for the development of channels along the cathode length was set due to the different initial electron concentration n_0 in each of the channels ($n_{01} = 10^4$, $n_{02} = 10^3$, $n_{03} = 10^2$, $n_{04} = 10$ cm^{-3}). The pumping conditions were the same as in experiments with the EOC. It follows from the calculations that at high energy inputs, oscillations will be observed in the glow of a separate diffuse channel, since the current through a separate channel is oscillatory, which qualitatively corresponds to the experimental results (Fig. 2.17).

If the simulation reduces the number of channels (keeping the cross-sectional area of a separate channel), then it is possible to observe modes when the amplitude of the channel current, the initial conductivity of which was minimal, significantly exceeds the current amplitudes in other channels (Fig. 2.27). This circumstance, due to the introduction of the phenomenological criterion of 'local critical current density', makes it possible, even in such a simple model, to qualitatively analyze the stability of SIVD.

It should be noted that the qualitative agreement between the results of calculations of the dynamics of the discharge channel structure and the experimental results of studying the discharge dynamics in the flat geometry of the electrodes and the blade-plane electrode system was obtained after the processes that could limit the current density in the channel were taken into account, namely: electron–ion recombination and electron impact dissociation of SF$_6$ molecules. A similar effect (the appearance of current oscillations in individual current channels of the SIVD) could be obtained by taking into account the adherence to vibrationally excited SF$_6$ molecules,

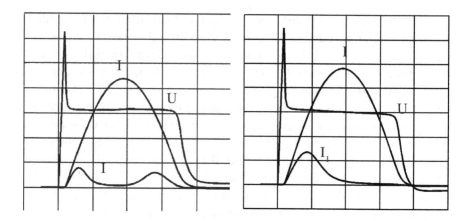

Fig. 2.26. Calculated oscillograms of voltage U, total discharge current I and current through the first channel I1: a) current 250 A / div., voltage 5 kV / div., sweep 100 ns / div; b) current 125 A / div., voltage 5 kV / div., sweep 100 ns / div.

in the same way as was done in [109, 114]. However, this requires assuming that a significant fraction of the energy deposited in the discharge goes to the excitation of the vibrational levels of the molecule, which contradicts the data [115, 116]. Chapter 4 is devoted to the study of the influence of vibrationally excited SF_6 molecules on the characteristics of the SIVD, anticipating these results, we note immediately that the role of the electron attachment process to vibrationally excited SF_6 molecules is not essential when analyzing the current limiting mechanisms of the diffuse channel.

Numerical simulation of the discharge of a capacitor for a discharge gap and comparison of calculated oscillograms with experimental ones shows that, within the limits of accuracy in estimating the k_d and β_{ei} constants, electron-separation processes by electron impact and electron–ion recombination almost compensate each other. Therefore, the main mechanism responsible for limiting the current density in the SIVD channel is the dissociation of SF_6 molecules and other components of the HF(DF) laser mixture. It is also obvious that in order to eliminate the effects that are difficult to interpret in an experiment due to the vibrational dynamics of conduction in the channel, it is desirable to create in the experiment such conditions so that it is possible to study the electrophysical characteristics of a single channel, i.e. create conditions under which new channels are not formed.

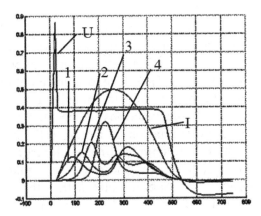

Fig. 2.27. Calculated oscillograms of voltage U, total discharge current I, and currents through individual channels 1–4 (time scale 100 ns/div).

It should be noted that this model (as, indeed, any 0-dimensional model that does not take into account the change in the volume of SIVD over time) claims, basically, a qualitative illustration of the process of redistributing currents in channels when considering possible mechanisms for limiting current density.

We also note that the separation of electrons from negative ions by electron impact must be taken into account when analyzing the processes determining the development of instability of a volume discharge in SF_6 and other strongly electronegative gases.

2.3. Qualitative analysis of the mechanisms for limiting the current density in the diffuse channel in SF_6

As already noted, the process of sequential formation of diffuse channels with a simultaneous decrease in current through previously generated during the ignition of a SIVD in a discharge gap with a flat geometry is largely determined by the mechanisms for limiting the current density in the conducting channel, depending on the specific energy input. Considering in this connection, following [69], some of the possible mechanisms. Among the various gases (SF_6, O_2, air, CO_2, N_2, He, Ar, Ne, C_3H_8, CO, C_2H_6) used in the above experiments, effects that can be associated with the existence of mechanisms for limiting the current density in the diffuse channel were observed only in highly electronegative gases [104].

First of all, we note that with E/N values close to the critical value, almost all the energy introduced into SF_6 goes to its dissociation [49]. The dominant channel is $SF_6 \rightarrow SF_4 + 2F$ [50]. The ionization potential of F atoms (17.4 eV) exceeds the corresponding value for SF_6 (15.7 eV) [117]. Therefore, a significant contribution of the F atoms formed to the total ionization should not be expected even with its significant concentrations. According to any of the possible mechanisms, the formation of negative F-ions obviously cannot compete with the attachment of the electrons to the SF_6 molecules [118]. The excitation threshold of F is equal to 12.7 eV [117] (the component of the main doublet with a threshold energy of ~0.05 eV can be disregarded in this case), i.e. lies in the region of intense excitation of SF_6 terms by electron impact [116]. In addition, there is reason to believe that inelastic processes with the participation of SF_4 molecules also do not have any significant effect on the energy spectrum of electrons in the discharge. Thus, bearing in mind the qualitative, basically, character of the models under consideration and neglecting in this connection the difference in the elastic scattering cross sections of electrons on SF_6 molecules and decomposition products, it can be assumed that the effect of the SF_6 dissociation affects only the reduction of the reduced field strength E/N by as the total concentration of neutral particles N increases with an increase in the specific energy input. It is reasonable to take the dependences of the main transport coefficients on E/N, as in pure SF_6. In particular, for an effective ionization coefficient we take $\alpha_{eff} = A \cdot N \cdot [E/N - (E/N)_{cr}]$. The total concentration of neutral particles N, depending on the specific energy introduced into the gas

$\int_0^t Ejd\tau$, is written in the form $N(t) = N_0(1 + \alpha \int_0^t Ejd\tau)$ Here, j is the current density, N_0 is the initial gas concentration. The coefficient α is expressed in terms of the dissociation constant k_d and is subsequently considered constant. The admissibility condition for such an entry is $-\alpha \int_0^t Ejd\tau \ll 1$. Turning to the 'energy' variable $q = \alpha \int_0^t Ejd\tau$ [69, 119], for the electron density $n_e(q)$ in the channel we have:

$$dn_e/d_q = (A/\alpha e) [1 - (1 + q)/K], \qquad (2.16)$$

where e is the electron charge, $K = E/E_{cr}$ is the overvoltage coefficient, E_{cr} is the critical field value for $N = N_0$. Integrating equation (2.16) we get:

$$n_e(q) = n_0 + \frac{B}{K}\left((K-1)q - \frac{q^2}{2}\right).$$ (2.17)

From equation (2.17) it follows that the dependence $n_e(q)$ has a maximum at $q = q_{max} = K-1$, and, therefore, as a result of SF6 dissociation, the electron density in the channel begins to fall at $q > q_{max}$. The current density in the channel $j(q) \sim n_e(q) \cdot (1 + q)^{-0.6}$, if we use the approximation of the dependence of the electron drift velocity in SF_6 on the E/N value from [13]. Obviously, the current density in the channel also passes through a maximum, but for $q = q' < q_{max}$. Under the experimental conditions, $K \approx 1$ and, consequently, $q \ll 1$, so the use of the approximation linear in q is justified. A significant argument in favour of the important role of the dissociation of SF_6 gas in SIVD is the fact that, according to [49], the fraction of the energy introduced into the discharge, which is used to decompose SF_6, increases with increasing E/N in the range of $E/N > 200$ Td. It can be verified that the inclusion of additional factors, significantly complicating the mathematical description, will not change the result obtained in principle – the dissociation of SF_6 by electron impact can be the main mechanism for limiting the current in the conducting channel in HF laser working media. This is confirmed by the results of numerical simulation of the discharge in SF_6, taking into account the process of SF_6 dissociation by electron impact (Section 2.2.2).

At a qualitative level, in [20, 69], using a similar approach, the influence of the attachment of electrons to vibrationally excited SF_6 molecules was considered. From this qualitative consideration, the attachment of electrons to vibrationally excited SF_6 molecules can also theoretically be a mechanism for limiting the current in a conducting channel in the working media of a non-chain HF laser. The results of studies presented in Chapter 4, however, show that, in fact, under conditions typical of non-chain electric-discharge HF laser, the role of this process in the effect of limiting current density is negligible.

Conclusions to chapter 2

1. In SF_6 and mixtures based on it, the SSVD develops in the form of SIVD (including in the discharge gap with high edge electric field amplification). For all its characteristics (appearance, current and voltage oscillograms), the SIVD in SF_6 does not differ from the volume discharge obtained using UV or X-ray preionization. It has

been established that the volumetric emission of the SIVD plasma is formed due to the overlapping of a large number of diffuse channels germinating from the cathode spots.

2. The dynamics of the formation of SIVD is as follows: when the voltage across the discharge gap reaches a higher breakdown in the zone of maximum amplification of the electric field, or in the region illuminated by UV radiation, one or several diffuse channels attached to the capacitance are formed, then the channels formed first initiate the appearance following diffuse channels. Thus, SIVD spreads through the discharge gap, perpendicular to the applied electric field, at a constant voltage close to the SF_6 critical voltage, gradually filling the entire discharge volume. The total number of diffuse channels increases in proportion to the energy introduced into the SIVD plasma.

3. It has been established that the use of cathodes with a rough surface (dimensions of asperities $50 \div 100$ μm) and the use of SF_6 gas mixtures with hydrocarbons, instead of $SF_6:H_2$, allows to accelerate the formation of new diffuse channels, as a result of which the plasma filling rate of the discharge volume increases and more homogeneous and steady discharge.

4. A simple numerical model has been created for calculating the characteristics of the SIVD in the working mixtures of non-chain HF (DF) laser, which also makes it possible to qualitatively study the dynamics of the formation of the SIVD.

5. The results of experimental studies of the dynamics of the SIVD development indicate the existence in SF_6 of mechanisms for limiting the current density in the diffuse channel, thanks to which, on cathodes with a rough surface, a uniform filling of large discharge volumes by the SIVD plasma occurs spontaneously without using special preionization devices.

3

Effect of limiting current density in the diffusion channel

In chapter 2, the results of the study of SIVD in SF_6-based gas mixtures were presented, from which it follows that the discharge has a fundamentally jet structure (represents a set of overlapping diffuse channels), and the combination of the observed features of the development of SIVD is due to the existence of processes that limit the current in a separate channel (current density restrictions). For a deeper understanding of the mechanisms responsible for the effect of limiting the current density, it was necessary to study the electrical characteristics of a single channel, i.e. create conditions under which new channels are not formed. This chapter presents the results of studies of the plasma of a single diffuse channel in SF_6 and SF_6-based mixtures with hydrocarbons. Mechanisms are studied that allow to obtain SIVD in working mixtures of HF(DF) lasers without using special pre-ionization devices. The general regularities of the formation of SIVD in SF_6 and mixtures based on it are discussed. The main results presented in this chapter are published in [69, 100, 119–121, 129].

3.1. Study of the characteristics of a single diffuse channel in SF_6 and mixtures based on it

3.1.1. Description of experimental installations and experimental techniques

The discharge chamber was a section of a cylindrical pipe 80 cm long, closed at both ends with transparent plexiglass flanges. The electrode system was mounted inside the discharge chamber near one of the flanges, which allowed visual inspection of the discharge uniformity. The SIVD was ignited in SF_6 and SF_6 mixtures with C_2H_6 and H_2 in the electrode geometry; the rod (cathode) is a plane with an interelectrode distance of d = 4–53 mm and a pressure of gas mixtures of p = 9–30 Torr. The cathode was a copper wire 1 mm in diameter in polyethylene insulation, in a discharge gap (DG) with such a cathode SIVD develops as a diffuse channel, closing on a single cathode spot [69, 119–121]. In order to fix the volume V_{dis} occupied by the SIVD plasma, the transverse dimensions of the discharge were limited to a dielectric tube of ~6–8 mm diameter. This made it possible to achieve large specific energy inputs (up to 1 J/cm^3), at which the influence of the mechanisms for limiting the current density becomes noticeable on the oscillograms of the voltage and current of the SIVD. A technique for studying a single diffuse channel in SF_6 and SF_6-based mixtures under conditions where the transverse dimensions of the channel were limited by a dielectric tube was proposed in [69] (the limited discharge method). Experiments with limited SIVD were also carried out with a cathode representing a metal rod 3 mm in diameter, sharpened ot a cone (the angle at the apex of the cone ~40°). Electric circuits and schematic representation of the discharge gap are shown in Fig. 3.1. At the discharge gap, the inductance L which varied during the experiments the pumping capacity C = 0.8÷8 nF, charged to a voltage of U = 13÷50 was discharged. Photographs of the SIVD made it possible to estimate the volume occupied by the diffuse channel plasma. Registration of the voltage and current of SIVD was carried out using respectively a calibrated voltage divider and a shunt on a digital oscilloscope with a 100 MHz bandwidth.

The value of the specific energy deposition in the SIVD plasma per one molecule, w_{in}, was determined as

$$. \, w_{in} = \frac{1}{V_{dis} \cdot N_0} \int_0^{t_{max}} U(t) \cdot I(t) \cdot dt. \tag{3.1}$$

Here N_0 is the initial total concentration of molecules, $U(t)$ is the voltage across the SIVD plasma, $I(t)$ is the SIVD current, t_{max} is the time when the current reaches its maximum (the duration of the leading edge of the current pulse). The time is counted from the moment of breakdown of the discharge gap. The value was measured at the time corresponding to the maximum current, $U_{qs} = U(t_{max})$, respectively, and integration in (3.1) is made to the value t_{max}.

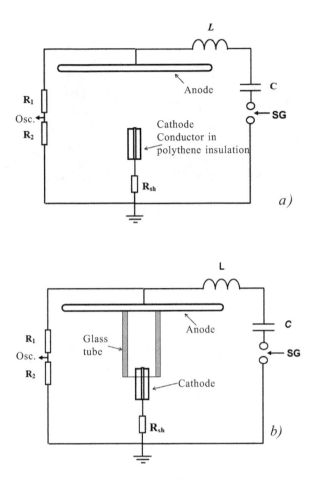

Fig. 3.1. Sxperimental installations for the study of the characteristics of free (*a*) and limited discharge (*b*). R_{sh} is a low-impedance shunt, SG is a controlled discharger, R_1 and R_2 are high-voltage voltage dividers.

The change in the magnitude with increasing win was characterized by the parameter $\Delta = (U_{qs}-U^0_{qs})/U^0_{qs}$, where U^0_{qs} is the value at the minimum level of the energy deposition in the plasma. The value was determined by experimental dependence $U^0_{qs}(w_{in})$ at $w_{in} \to 0$.

The critical value of the reduced electric field strength $(E/N)_{cr}$, corresponding to the value (in strongly electronegative gases, the value $(E/N)_{cr}$ is close to the value of E/N in the quasistationary phase of the discharge [99]), was determined from the relation:

a   *b*

Fig, 3.2. Photos of SIVD in the system of rod–plane electrodes, with $d = 4$ cm, $p = 33$ mmHg. in a mixture of $SF_6 : C_2H_6 = 10: 1$: *a*) $W = 0.73$ J; *b*) $W = 2.33$ J.

$$U_{qs}^0 = (E/N)_{cr.} d \cdot N_0 + U_c, \qquad (3.2).$$

where d is the interelectrode distance, U_c is the cathode potential drop; the thickness of the cathode layer in the pressure range under consideration can be neglected [45]. To this end, in the studied gas mixtures, dependences of U_{qs}^0 on the parameter pd were taken at low values of the specific energy input, $w_{in} < 0.05$ eV/molecule. The pressure of the mixtures varied in the range $p = 6$–60 Tor. In this series of measurements, SIVD was not limited to a dielectric tube. Dependences $U_{qs}^0 = f(pd)$ for all gas mixtures are approximated with good accuracy by linear functions.

The study of the stability of SIVD in SF_6 and mixtures based on it in the electrode system shown in Fig. 3.1 *a* was carried out according to the method described in the second chapter of this book (Section 2.1.4).

3.1.2. Investigation of the characteristics of a single diffuse channel unlimited by external walls

Figure 3.2 presents photographs of SIVD in the rod–plane electrode system (Fig. 3.1 *a*) with $d = 4$ cm, $p = 33$ mmHg in an $SF_6 : C_2H_6 = 10:1$ mixture, taken at two values of the energy W introduced into the discharge plasma. In this geometry, the SIVD is a diffuse channel expanding towards the anode with a bright plasma formation at the cathode, and, as can be seen from Fig. 3.2, as W increases, the volume V of the diffusion channel increases. Attention is drawn to the high stability of the discharge in the considered discharge gap geometry, despite the fact that the current density at the cathode reaches $j = 1.2 \times 10^5$ A/cm^2.

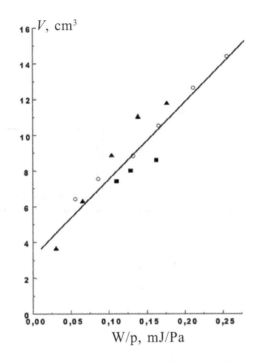

Fig. 3.3. Dependence of the volume V occupied by the discharge on the parameter W/p for a mixture of the composition $SF_6:C_2H_6 = 10:1$: ■) $p = 15.5$ mmHg. ▲) $p = 33$ mmHg; O) $p = 49.5$ mmHg.

The maximum value of the volume V occupied by the SIVD by the time of the end of the energy input, with the same values of the total energy W introduced into the plasma, increased with decreasing pressure of the mixture p. Figure 3.3 shows the dependence of V on the parameter W/p, taken at different values of p. It is seen from Fig. 3.3 that this dependence is satisfactorily described by a linear function within the accuracy of determining V in the experiment. This, apparently, reflects the fact that, starting with a certain value $V = V_0$, determined by the stage of the initial rapid expansion of the channel, a further increase in volume occurs at an approximately constant value of the parameter $\Delta W/(p \cdot \Delta V) \approx const \approx 2.1 \cdot 10^{-2}$ (mJ / (cm³·Pa)), where $\Delta V = V - V_0$ is the volume increase, ΔW is the energy introduced into the volume ΔV.

The dependences of V on W, obtained for pure SF_6 and a mixture of SF_6 with ethane, are presented in Fig. 3.4. As can be seen from this figure, a linear increase in V is observed with increasing W (the dependences are obtained at constant pressure). At $W = 3.5$ J, the value of V reaches ~18 cm³, and the transverse size of the

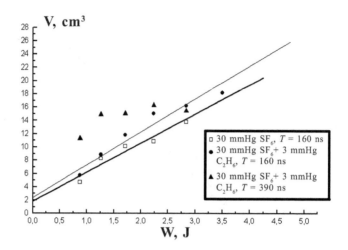

Fig. 3.4. Dependences of the volume V occupied by the diffuse channel on the energy W deposited in the SIVD plasma, obtained in pure SF_6 and an SF_6/C_2H_6 mixture with a discharge current duration $T = 160$ and $T = 390$ ns.

channel at the anode is ~3.5 cm. In this geometry, the influence of the current duration on the value of the volume occupied by the discharge is noticeable. A similar effect was observed in experiments with flat electrodes (section 2.1 of this book). It is noteworthy that the volumes occupied by the SIVD in pure SF_6 and SF_6 mixtures with ethane, for identical values of the total energy introduced into the plasma W do not differ significantly (Fig. 3.4), and the relative increase in the limiting energy contribution to the discharge plasma for a given duration (increased stability) by adding C_6H_6 to SF_6, it is almost the same as in the gap with flat electrodes.

Figure 3.5 shows the dependences of the limiting value of the parameter $CU^2/2$ on $T = \sqrt{LC}$ for various gas mixtures in the rod-plane system of electrodes.

A comparison of Fig. 3.5 with Fig. 2.10 *a* (flat electrodes) shows that with the same distance between the electrodes in the rod–the plane gap the addition of ethane to SF_6 increases the marginal energy input in approximately the same proportion as in the geometry of flat electrodes. The C_2HCl_3 additive has a harmful effect on the stability of SIVD in SF_6, despite the fact that, in the system of flat electrodes, the addition of this substance increases the cathode spot density.

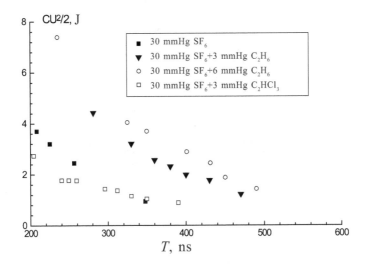

Fig. 3.5. Dependences of the parameter $CU^2/2$ on the contour parameter $T = \pi\sqrt{LC}$ for various gas mixtures in the electrode–needle–plane system.

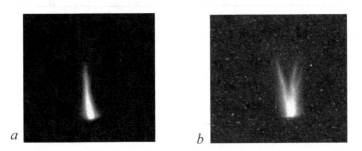

Fig. 3.6. Photos of the cathode torch: a) in pure SF_6, $P = 30$ Torr; b) SF_6: $C_2H_6 =$ 10: 2 mixture with a total pressure of 36 Torr.

Thus, the addition of C_2H_6 to SF_6 equally increases the stability of the SIVD both in the system of flat electrodes, where the density of the cathode spots increases significantly in the mixture of SF_6 and C_2H_6, and in the electrode–rod–plane system, where the SIVD develops from one cathode spot. The detected effect is clarified by a photo of a cathode plume, i.e. bright plasma channel growing directly from the cathode. Figure 3.6 presents photographs of the cathode plume obtained with the same energy input in pure SF_6 (Fig. 3.6*a*) and SF_6:$C_2H_6 = 10$:2 mixture (Fig. 3.6 *b*) (diffuse luminescence was attenuated using light filters, therefore it is not visible in the photo). For Fig. 3.6 *b*, all linear dimensions were increased by 2

U = 5 kV/div, I = 145 A/div

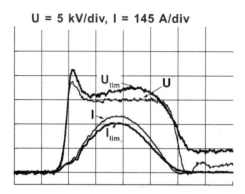

Fig. 3.7. Oscillograms of SIVD voltage and current limited (U_{lim}, I_{lim}) and unlimited (U, I) in an SF_6:C_2H_6 = 10:1 mixture at P = 20 mmHg and d = 5.8 cm. Sweep 50 ns/div.

times, since the size of the plume in the SF_6:C_2H_6 = 10:2 mixture was significantly less than in pure SF_6. It can be seen that in pure SF_6 the cathode plume has the shape of a cone strongly elongated towards the anode, while in a mixture with ethane the cathode plume has several cone-shaped tops in its upper part (3 vertices can be distinguished in this photo). It turns out that the formation on the body of a bright plasma channel of new peaks is like an analogue of the formation of new cathode spot in the system of flat electrodes. However, it is obvious that the formation of such vertices is associated with the processes occurring in the channel plasma. Therefore, it can be assumed that the increase in stability in this case is due to a change in the gas-discharge kinetics.

3.1.3. Investigation of the characteristics of the diffuse channel bounded by external walls

Figure 3.7 shows oscillograms of limited (U_{lim}, I_{lim}) and unlimited (U, I) voltage and current of the SIVD in an SF_6:C_2H_6 = 10:1 mixture at P = 22 mmHg. and d = 5.8 cm. It is clear from this figure that when the SIVD volume was not limited to a tube, the oscillograms of voltage and current in the rod-to-plane system are no different from typical oscillograms obtained in the system of flat electrodes. Limiting the size of the discharge leads to a noticeable change in the type of waveforms. In Fig. 3.7 it is especially noticeable that in general, the limited discharge voltage is higher than the unlimited (U_{lim}> U), and the discharge current is less (I_{lim} < I).

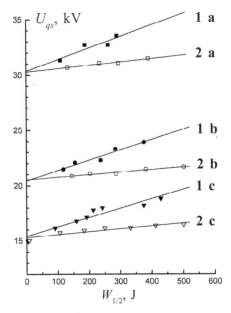

Fig. 3.8. Dependences of U_{qs} on the energy $W_{1/2}$, for limited (1) and unlimited (2) SIVD. A mixture of $SF_6:C_2H_6 = 10:1$, $d = 4$ cm: 1c, 2c – $P = 33.6$ mmHg; 1b, 2b – $P = 45.8$ mmHg; 1a, 2a – $P = 67.2$ mmHg.

Thus, in the experiment considered here, an extremely unusual situation is observed for independent discharges: the plasma conductivity decreases when the discharge volume is limited, i.e. with an increase in the specific energy input. We note that the modification of the voltage pulse observed in experiments with limiting the size of the SIVD is associated precisely with an increase in the specific energy input to the plasma, since the discharge cannot expand under the conditions of this experiment during energy input.

Figure 3.8 presents the dependences of U_{qs} on the energy $W_{1/2}$ introduced into the discharge plasma up to the current maximum in a rod–plane system of electrodes for limited and unlimited SIVD. The value of U_{qs} was determined at the maximum current. From Fig. 3.8 it can be seen that with increasing $W_{1/2}$ in the limited SIVD, a noticeable increase in U_{qs} is observed, significantly more pronounced than in the unlimited SIVD. The relative increase in U_{qs} (in comparison with U_{qs} of unlimited SIVD) with the same $W_{1/2}$ is greater the smaller the pressure, i.e. the more energy is introduced into the plasma per unit volume and per molecule. The results presented in Figs. 3.7 and 3.8 indicate a decrease in the effective

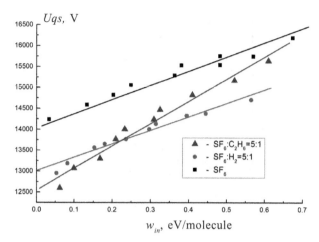

Fig. 3.9. Dependences of U_{qs} on w_{in} at t_{max} = 30 ns and pressure p = 24 Torr in SF_6 (■), as well as in mixtures of SF_6:C_2H_6 = 5:1 (▲) and SF_6:H_2 = 5:1 (●).

ionization coefficient in mixtures based on SF_6 with an increase in the specific energy input to the discharge plasma.

When analyzing the elementary processes occurring in the plasma of a single SIVD channel (SSVD), it is much more convenient to analyze the dependencees similar to those shown in Fig. 3.8, not on the absolute energy, but on the specific energy introduced to the current maximum.

Figure 3.9 shows the dependences in a limited discharge on the value of w_{in} in SF_6 and SF_6 mixtures with C_2H_6 and H_2. In the range of changes w_{in} = 0 ÷ 1.5 eV/molecule these dependences can be approximated with a fairly good accuracy by a linear function.

$$U_{qs} = U_{qs}^0 + K \cdot w_{in}, \qquad (3.3)$$

where K is a constant depending on the composition of the gas mixture. A noticeable increase of U_{qs} with an increase in the energy introduced into the plasma directly indicates the existence of mechanisms for limiting the current density. In the experiments described, a single diffuse SIVD channel is modelled, the expansion possibilities of which are limited by the walls of the dielectric tube. In the absence of a tube, the volume occupied by the discharge will increase with an increase in the energy input due to the current density limiting mechanisms, and in the case of the flat geometry of the electrodes, these mechanisms will prevent the flow of all the energy stored in the capacitor through one channel and stimulate diffuse channels [69, 121]. As already noted, the most likely mechanism for limiting the current density in the diffuse channel is the dissociation of neutral molecules by electron impact. In this connection, the question of the quantitative assessment of the effectiveness of this process in the conditions under consideration is of current importance. Below is a theoretical justification of the choice of criteria for the quantitative assessment of the contribution of the process of diisociation of mixture molecules by electron impact, first proposed in [120, 121].

Let the mixture consist of components, each of which admits nl different dissociation channels, and as a result of one dissociation act of the l-th component, fragments are formed in the i-th channel. Then for the rate of increase in time of the total concentration of dissociation products in this channel, we have:

$$\frac{dN_d^{il}}{dt} = \frac{1}{e} m_{il} \eta_d^{il}(E/N) JE \quad 1 \leq i \leq n^l \; 1 \leq l \leq S \tag{3.4}$$

$$\eta_d^{il}(E/N) = \frac{k_d^{il}(E/N)}{u_e(E/N)} \frac{\delta_l}{(E/N)} \quad \delta_l = \frac{N_l}{N} \tag{3.5}$$

Here e is the electron charge, J is the current density, N_l is the concentration of the l-th component; $k_d^{il}(E/N)$ and $u_e(E/N)$ are the constant of the rate of dissociation by the electron impact of the l-th component in the i-th channel dependent on the reduced strength of the electric field E/N, and the drift velocity of the electrons, respectively. The value introduced by us has a simple physical meaning. Indeed, as follows from eq. (3.4), this is the energy value of one act of dissociation of the l-th component in the i-th channel. The total concentration of all dissociation fragments formed by the time the current reaches its maximum is equal to

$$. \; \Delta N_d = \frac{1}{e} \sum_l \sum_i m_{il} \int_0^{t_{max}} \eta_d^{il}(E/N) JE dt . \tag{3.6}$$

When writing (3.6), it was taken into account that the total number of fragments of molecules formed during dissociation persists during the discharge pulse.

With relatively small specific energy inputs, the reduced field strength E/N varies little as the discharge current increases. In addition, the numerator and denominator in relation (3.5) are monotonously increasing functions of the parameter, and as a result, the magnitude $\eta_d^{il}(E/N)$ for the characteristic conditions of our experiments (energy is introduced into the plasma in the quasistationary discharge phase at close to a constant value) is weakly dependent on E/N. In particular, using the calculated data on the kinetic coefficients in SF_6, it is easy to see that $\eta_d^{il}(E/N)$ in pure SF_6 it increases by less than 10% even with a fivefold increase in the reduced field strength from $100Td$ to $500Td$ [121]. This allows us to put in (3.6) the sign of the integral. As a result, taking into account (3.1), we have:

$$\Delta N_d = \frac{1}{e} w_{in} N_0 \sum_l \sum_i m_{il} \eta_d^{il}(E/N). \tag{3.7}$$

We define the degree of dissociation α as the total excess compared with the initial density N_0, the concentration of particles in the dissociated gas $\Delta \tilde{N}_d$, referred to N_0. Since in each dissociation event one molecule of the original component disappears,

$$\Delta \tilde{N}_d = \frac{1}{e} w_{in} N_0 \sum_l \sum_i (m_{il} - 1)\eta_d^{il}(E/N) \quad \alpha = \frac{\Delta \tilde{N}_d}{N_0}, \tag{3.8}$$

The energy value of the formation of one of the fragments of the initial components of the mixture, regardless of their nature, is

$$E_d = \frac{w_{in} N_0}{\Delta N_d} = \frac{e}{\sum_l \sum_i m_{il} \eta_d^{il}(E/N)}. \tag{3.9}$$

Values E_d and α are interrelated by the obvious relationship:

$$E_d = \frac{w_{in}}{\alpha} \xi, , \quad \xi = 1 - \frac{\sum_l \sum_i \eta_d^{il}}{\sum_l \sum_i m_{il} \eta_d^{il}} \tag{3.10}$$

To determine parameter ξ, it is necessary to have sufficiently complete information about all dissociation channels, which is currently hardly available for mixtures of polyatomic molecules. At the same time, it is for mixtures of polyatomic gases that one can expect that, as follows from (3.10), ξ differs little from unity.

Let us express E_d in terms of the quantities measured in the experiment, assuming that the increase in the combustion voltage of SIVD with an increase in the energy input is due exclusively to dissociation. In accordance with (3.2) and (3.3) we have:

$$(E/N)dN_0(1+\alpha)+U_c = U_{qs}^0 + Kw_{in} \tag{3.11}$$

When discharging in a dissociating gas, the reduced field strength may differ from its critical value (E/N) in the absence of dissociation: $E / N = (E / N)_{cr}(1+\varepsilon)$. At low degrees of dissociation ($\alpha \ll 1$), this difference, however, is small and it can be assumed without special error that $|\varepsilon| \ll 1$. With this in mind, from (3.2), (3.11) it follows:

$$(E / N)_{cr}(\varepsilon + \alpha)dN_0 = Kw_{in}. \qquad (3.12)$$

The obtained experimental data do not allow to determine ε either by sign or by absolute value. One can only talk about a rough estimate α or E_a/ξ. In particular, neglecting ε compared with α, we have:

$$w_{in} / \alpha = (E / N)_{cr} dN_0 / K. \qquad (3.13)$$

The values of w_{in}/α, determined with regard to this assumption by experimental dependences, are presented in Table 3.1. Despite the indicated inaccuracy in determining the value of w_{in}/α, the results given in Table 3.1 allow us to make a number of conclusions of a fundamental nature. First, the degree of dissociation α can, with a sufficiently noticeable specific energy input (Fig. 3.10), exceed 10%. In other words, in the conditions under consideration, the process of dissociation of the components of the studied mixtures by electron impact does indeed proceed rather effectively. Secondly, it can be seen that the value w_{in}/α strongly depends on the composition of the gas. In mixtures of SF_6 with ethane, it is significantly lower than in pure SF_6 or mixtures of SF_6 with hydrogen. The latter is apparently due to the fact that during dissociation, the C_2H_6 molecule can decompose into more than two fragments (relation (3.8)). In addition, the dissociation constant of the H_2 molecule by electron impact is relatively small, since the total dissociation cross section reaches its maximum value of $\sigma_{dis} \approx 8 \cdot 10^{-17}$ cm^2 only at an electron energy of ~15 eV, and the effective mechanism of dissociation through the excitation of vibrational levels does not have time to 'turn on' during

Table 3.1. Values of parameter w_{in}/α for SF6 and SF_6:H_2 and SF_6:C_2H_6 mixtures

Mixture	w_{in}/α, eV/mol
Pure SF_6	5 ± 1
SF_6:H_2 = 5:1	5.2 ± 1.1
SF_6:C_2H_6 = 5:1	3.1 ± 0.9
SF_6:C_2H_6 = 5:5	2.6 ± 0.6

the discharge [122]. With an increase in the percentage of ethane (increase of δ_l), the value of w_{in}/α, in accordance with (3.8), should decrease, since the dissociation constant for ethane in the conditions under consideration is certainly higher than for SF_6. This is in full compliance with the data given in Table 3.1.

The following circumstance justifies to a large extent the assumptions made in assessing α. According to [50], under SSVD conditions, the dissociation channel dominates in pure SF_6

$$SF_6 + e \rightarrow SF_4 + 2F + e. \tag{3.14}$$

In [50], for these conditions, the energy value of the formation of a fluorine atom $\tilde{E}_d(F)$ was experimentally determined, equal to the ratio of the specific energy input $w_{in}N_0$ to the total concentration of fluorine atoms $\Delta \tilde{N}_d(F)$ formed during the dissociation process. Since, in accordance with the reaction (3.14) $\Delta \tilde{N}_d(F) = \Delta \tilde{N}_d$, then, taking into account the relation (3.8), we have:

$$\tilde{E}_d(F) = \frac{w_{in}N_0}{\Delta \tilde{N}_d(F)} = \frac{w_{in}N_0}{\Delta \tilde{N}_d} = \frac{w_{in}}{\alpha} \tag{3.15}$$

In [50] and [51], the values $\tilde{E}_d(F) = 4^{+12}_{-0.8}$ eV and 4.5 eV, respectively, are obtained, which is in very good agreement with our result $w_{in}/\alpha = 5\pm1$ eV (Table 3.1). This indirectly confirms the dominant role of the dissociation reaction (3.14) in pure SF_6. Note that the energy value for the formation of one fragment in the reaction (5.14) is somewhat lower. Indeed, in this case $S = 1$, $m_{il} = 3$ $\xi = 2/3$. As a result, using Table 3.1, from the relation (3.10) we get $E_d = 3.3$ eV.

Due to the lack of detailed information on the dissociation channels for the mixtures of SF_6 with C_2H_6, the values w_{in}/α given in Table 3.1. can no longer be used to determine the cost of formation of fluorine atoms. However, they give a reasonable idea of some of the average energy costs for the formation of a dissociation fragment, regardless of its nature.

Figure 3.10 shows the dependences of the parameter $\Delta = (U_{qs} - U_{qs}^0)/U_{qs}^0$ on w_{in}, constructed according to the data of Fig. 3.9. It can be seen that the parameter Δ in the SF_6:C_2H_6 mixture is larger in general and grows with an increase in w_{in} noticeably faster than in SF_6 and the SF_6:H_2 mixture Within the limits of linear approximation accuracy (3.3)

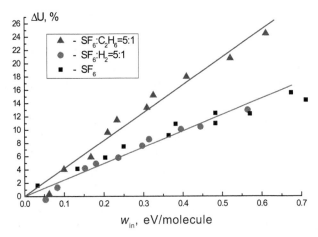

Fig. 3.10. Dependences of the parameter $\Delta = (U_{qs} - U_{qs}^0) / U_{qs}^0$ on w_{in} at pressures of $p = 24$ Torr and $= 30$ ns in SF_6 (∎), as well as in SF_6:C2H6 $= 5:1$ (▲) and SF_6: H2 $= 5: 1$ mixtures (●).

$$\Delta = \frac{K}{U_{qs}^0} w_{in}, \qquad (3.16)$$

that is, the value of Δ increases in direct proportion to w_{in}. In these experiments, it was also found that in the SF_6 mixtures with hydrocarbons at constant w_{in} the value of Δ increases with increasing percentage of hydrocarbon in the mixture (but not more than by 50%), and the increase (decrease) of the parameter Δ correlates with the increase (decrease) of stability of the SIVD.

An obvious correlation between the Δ and w_{in}/α values should also be noted: with the same w_{in} Δ is the larger the smaller the w_{in}/α, i.e., the relative increase in the burning voltage of the SSVD, which is related to the limitation of the current density, the greater the dissociation of the mixture in the SSVD plasma. In fact, the parameter Δ serves as an indicator of the very possibility of the ignition of the SSVD in the form of SIVD and determines, all other conditions being equal, its stability margin.

Note that the linearity of the experimental dependences of U_q and Δ on w_{in}, taking into account the above theoretical consideration, as well as the good agreement of the values of the energy value for the formation of a fluorine atom determined in this work with the values of this parameter obtained using other experimental methods [50, 51], directly indicate in favour of the assumption of the decisive role of the process of dissociation of SF_6 and other components of the HF laser working medium by electron impact in the effect of limiting the current density.

Thus, from the presented experimental material it can be concluded that in SF_6 mixtures with C_2H_6, the effect of limiting the current density associated with the dissociation of molecules in the SIVD plasma is more pronounced than in pure SF_6 and SF_6 mixtures with hydrogen. That is why with the development of SIVD in SF_6 mixtures with hydrocarbons in systems with flat electrodes, new diffuse channels will appear at lower values of the electrical energy introduced into the SIVD plasma than in mixtures with a higher dissociation energy. As a result, the channels should become larger, i.e., the uniformity of the SIVD should increase, and due to a decrease in the current through a separate cathode spot, its stability should also increase (the time of steady burning will increase). Apparently, it is precisely these factors that are responsible for the higher stability and homogeneity of the SIVD in the SF_6 mixtures with hydrocarbons (carbon deuterides) as compared to the SIVD in the SF_6 mixtures with hydrogen (deuterium).

3.1.4. Numerical simulation of limited SSVD

The equivalent electrical circuit and the complete system of equations for modelling the SSVD are described in section 2.2 of the book. To illustrate how taking into account each elementary nonlinear process (dissociation, electron–ion recombination, electron detachment) appears on the oscillograms of voltage and current, Fig. 3.11a shows the oscillogram of the voltage of the limited SSVD in SF_6 at a pressure $p = 15$ Torr and $W_{el} = 0$, 12 J·cm^{-3}. Figure 3.11b shows the calculated oscillograms of voltage, upon receipt of which the model took into account the following processes: ionization by electron impact and electron attachment (IA); ion–ion recombination (IIR, rate constant $\beta_{ii} = 2 \cdot 10^{-8}$ cm^3 · s^{-1} [98]); dissociative electron–ion recombination (EIR, the value in the calculation varied in the range $\beta_{ii} = 2 \cdot 10^{-8}$ cm^3·s–1); detachment of electrons from negative ions by electron impact (DE, $k_d = 0.5–3 \cdot 10^{-7}$ cm^3 s^{-1}); SF_6 dissociation by electron impact (DEI, the cost of formation of a fluorine atom is ~4.5 eV). The rate constant for the detachment of electrons from a negative ion (DE), $k_d = 10^{-7}$ cm^3/s, was estimated under the assumption that it should not be less than the rate constant for elastic scattering of electrons by SF_6; $\beta_{ei} = 10^{-7}$ cm^3/s, estimated from the assumption that $\beta_{ei} \sim T_e^{-1/2}$. Comparison of the calculated and

81

experimental oscillograms of the voltage of the SSVD shows that their closest matching is achieved at $k_d \approx \beta_{ei}$. Thus, the dissociation of SF_6 in the SSVD plasma, as well as the dissociation of C_2H_6 in an SF_6:C_2H_6 mixture, is, apparently, the main mechanism for limiting the current density.

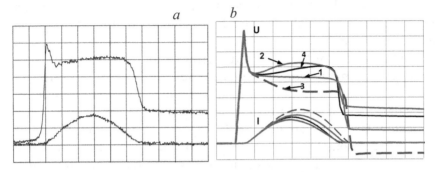

Fig. 3.11. Oscillograms of voltage U and current I of SSVD in SF_6 at $p = 15$ Torr and an interelectrode distance of 48 mm (limited discharge): *a*) experimental oscillograms, $U = 2$ kV/div, $I = 50$ A/div; *b*) – calculated oscillograms, $U = 2$ kV / div, $I = 50$ A/div. The calculation took into account the processes: 1 – EA; 2 – EA, EIR; 3 – EA, NI; 4 – EA, DE I with $kd = \beta_{ei}$.

3.1.5. Analysis of the results

It has been established that in the SF_6-based gas mixtures, including in working mixtures of the non-chain HF laser, the SSVD burning voltage increases almost linearly with the energy introduced into the plasma. Thus, the results presented in this section directly illustrate the existence of current density limiting mechanisms in the diffuse channel plasma in SF_6 and mixtures based on it. Assuming that the process of dissociation of molecules by electron impact plays the main role in increasing the combustion voltage of the SSVD with increasing input energy, approximate values of the energy value for the formation of dissociation fragments in the HF lasers were found from experimental dependences. The energy value of the formation of a fluorine atom in an electric discharge for pure SF_6 is determined. The value obtained in this section agrees well with the results of [50, 51] obtained by other methods.

It has been established that the addition of ethane to SF_6, in contrast to the addition of hydrogen, enhances the effect of limiting the current density and leads to a significant increase in the voltage growth rate on the SIVD plasma with an increase in the energy introduced into the plasma due to the dissociation of hydrocarbon molecules by electron impact under conditions of high E/N, determined by the characteristics of SF_6, as the main component in the mixture. This suggests that it is precisely the enhancement of the effect of limiting the current density that is the main reason for increasing the stability and homogeneity of the SIVD when hydrocarbons are added to SF_6.

It should be noted that the attraction of current limiting mechanisms in the conducting channel alone is not enough to fully understand the processes observed in the experiment, in particular, the discharge propagation deep into the gap in a direction perpendicular to the applied field. Apparently, radiation effects of the diffuse channel itself play a significant role in the transverse direction expansion effects of the diffuse luminescence zone of a single channel, as well as the propagation of a SIVD in a gap with a flat geometry, since hard UV radiation is present in the spectrum of this glow, and, possibly, soft X-rays radiation [123, 124]. Indeed, the formation of a discharge in such a strongly electronegative gas as SF_6 occurs at a high value of E/N, therefore, a significant part of the electrons can go into continuous acceleration mode, i.e. so-called runaway electrons appear [123–125].

3.2. Characteristics of the SSVD under conditions of strong population of vibrationally excited states of SF_6 molecules

The following processes were theoretically considered as possible mechanisms for limiting the current density in the previous sections of the thesis: SF_6 dissociation by electron impact, electron–ion recombination, and the attachment of electrons to vibrationally excited SF_6 molecules. The need to take into account the latter process when calculating the SSVD electrical characteristics in SF_6 was also noted in a number of papers [109–111]. However, special experiments that allowed determining the role of the attachment of electrons to vibrationally excited SF_6 molecules under SSVD (SIVD) conditions were not performed in these works. In a number of works [116, 126], the cross sections for the attachment of electrons to the vibrationally excited SF_6 molecules were measured, but, nevertheless, for an accurate and consistent calculation of the characteristics of the SSVD, data on the cross sections of various elementary processes occurring in the SSVD plasma in a wide range of E/N

is not enough. Therefore, it is advisable to obtain more reliable data on the effect of the vibrationally excited SF_6 molecules on the combustion voltage and other SSVD characteristics directly from the experiment. However, it should be noted that it is rather difficult to study the effect of the attachment of electrons to the vibrationally excited SF_6 molecules on the combustion voltage of the SSVD under conditions when the vibrationally excited molecules are produced in the discharge itself. Therefore, to study the effect of the vibrationally excited SF_6 molecules on the characteristics of SSVD in HF laser working mixtures, we proposed to populate the vibrational degrees of freedom of molecules by irradiating mixtures containing SF_6 by radiation from a pulsed CO_2 laser [127–129].

This section of the thesis presents the results of studies of the role played by the attachment of electrons to the vibrationally excited SF_6 molecules in the effect of limiting the current density.

3.2.1. Experimental setup and experimental methods

Studies of the effect of vibrational excitation of an SF_6 molecule on the characteristics of SSVD were carried out on an installation schematically depicted in Fig. 3.12. Vibrationally excited SF_6 molecules were obtained due to the resonant absorption of CO_2 laser radiation (line P20 of the 10.6 μm band) and the subsequent redistribution of the absorbed energy over the vibrational degrees of freedom of the SF_6 molecules during the V–V exchange. A laser beam with a transverse size of 60×60 mm with a uniform distribution of energy over the cross section was introduced into the discharge chamber through an input window of BaF_2. The discharge was photographed through another window with a digital camera (C). A

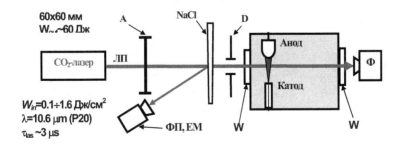

Fig. 3.12. The scheme of the experimental setup.

wedge of NaCl (or Ge) was placed in the path of the laser beam, with part of the radiation was directed to an IR photodetector (PD). In experiments to study the effect of CO_2 laser radiation on the structure of the discharge, various diaphragms (D) were installed in the path of the laser beam passing through the wedge. The density of the laser radiation at the entrance to the discharge chamber by installing calibrated attenuating films (A) in the path of the laser beam was smoothly controlled in the range $W_{in} = 1.5 \div 0.1$ J/cm^2.

The discharge chamber was a segment of a polyethylene pipe with an inner diameter of 20 cm and a length of 20 cm. From the ends, the chamber was sealed with flanges on which BaF$_2$ windows were attached. The chamber was filled with mixtures of SF$_6$ with various gases (C$_2$H$_6$, Ne, He, C$_3$F$_8$); the partial pressure of SF$_6$ in the experiments was 3–30 Torr. The cathode was a copper wire (diameter 1 mm), placed in a dielectric tube with an inner diameter of 1 mm and an outer 7 diameter of mm. The anode was a cylinder of Al 10 mm diameter and a length of 8 cm, the axis of the cylinder was perpendicular to the laser beam and the axis of the rod (cathode). The 'frontal plane of discharge' was an imaginary plane passing through the cylinder axis and the rod and was located at a distance of 25 mm from the entrance window (BaF$_2$) through which laser radiation was introduced into the chamber. Thus, the discharge thickness in the direction of the laser beam did not exceed 10 mm. This discharge gap geometry was chosen specifically to, firstly, to avoid the appearance of more than one diffuse channel and, secondly, to minimize the influence of the irregularity of absorption of CO_2-laser radiation in a gas.

The electrical circuit for obtaining the SSVD in SF$_6$ is depicted in Fig. 3.13. A capacitor with a capacity of $C = 1$–4 nF, charged to a voltage of $U = 20$–40 kV, through inductance L was discharged at the discharge gap with an interelectrode distance of $d = 40$–50 mm. The current and voltage in the discharge gap were recorded on a TDS-220 digital oscilloscope using a Rogowski coil and a resistive voltage divider, respectively. In the experiments on oscillograms of current and voltage on the SSVD plasma, the change in the electrical properties of the SSVD under variations in the energy density of the laser radiation absorbed in the discharge volume and the delay time between the arrival of the CO_2 laser pulse and the supply of a voltage pulse to the discharge chamber was studied.

Figure 3.14 shows typical oscillograms of a CO_2 laser pulse, obtained by an IR photodetector, and a voltage pulse on a discharge

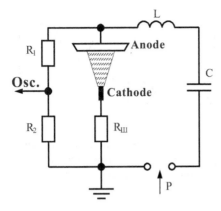

Fig. 3.13. Electrical circuit of the installation for the ignition of SSVD in a vibrationally excited gas.

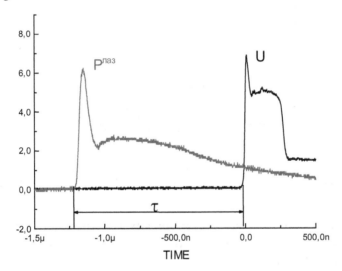

Fig. 3.14. Oscillograms of red laser radiation pulses and voltages on a discharge chamber U.

chamber. As can be seen from this figure, the voltage pulse on the SSVD plasma has a characteristic shape for highly electronegative gases – it clearly shows the quasistationary phase of the discharge (the shelf on the oscillogram). The voltage in the quasi-stationary phase – the so-called burning voltage – measured by oscillograms of voltage and current of the SSVD, as the magnitude of the voltage at the current maximum. A special experiment was conducted with gas

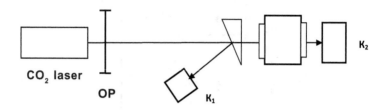

Fig. 3.15. An experimental setup for determining the density of the laser energy absorbed in the discharge volume. OP – Teflon films to attenuate radiation; K_1, K_2 – calorimeters for determining the incident and transmitted energy of the CO_2 laser; K – NaCl. wedge.

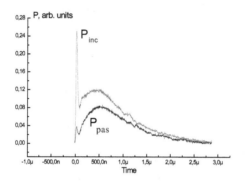

Fig. 3.16. Oscillograms of laser radiation pulses: P_{inc} – the pulse incident on the input window; P_{pas} – impulse that passed the cell with a thickness of 20 mm filled with SF_6, 15 Torr.

cells to determine the laser energy density absorbed in the discharge region, i.e. in a gas layer 10 mm thick, separated at a distance of 20 mm from the entrance window.

The scheme of the experiment is shown in Fig. 3.15. The cells were cylinders with a length of 20 and 30 mm, respectively; NaCl windows were attached to the ends. The energy of the laser radiation that passed through the cell was measured using a TPI2-5 calorimeter; in special experiments, an IR photodetector was installed instead of the calorimeter.

Figure 3.16 shows laser radiation pulses: incident on a cell and passing through a cell 20 mm thick, filled with SF_6 (15 Torr). As can be seen from this figure, the pulse shape of the CO_2 laser after passing through the gas layer is noticeably distorted. Despite the fact that the absorption cross sections of various CO_2 laser lines by an SF_6 molecule are known from the literature, calculating the absorption coefficient of mixtures based on SF_6 for a particular laser pulse is quite a difficult task, since it is necessary to take into account the change in the distribution of molecules over different vibrational rotational levels during a pulse. Therefore, the absorption coefficient of CO_2 laser radiation of various mixtures was determined experimentally. The experiments investigated the transmission of cells filled with different gas mixtures, depending on the density of the incident laser energy. The energy density incident on the cell was changed due to the installation of Teflon films in the ???. The laser energy absorbed in the discharge zone was determined as the difference (taking into account the loss on the windows) between the transmittance values of the cell with a length of 20 mm and 30 mm, respectively.

3.2.2. Determination of CO_2 laser energy absorbed in the discharge gap

The results of measurements of the transmittance of cells with a thickness of 20 mm and 30 mm are presented in Fig. 3.17. which shows for two different gas pressures the dependences of the transmittance of a layer of gas 20 mm and 30 mm thick on the density of the laser energy W_{in} incident on the gas. The transmittance T was determined by the formula $T = E_p/E_{in}$, where E_p is the energy of the CO_2 laser passed through the gas layer, E_{in} is the energy incident on the gas. After approximation of these experimental data, we plotted the dependences of the energy W_a absorbed in a

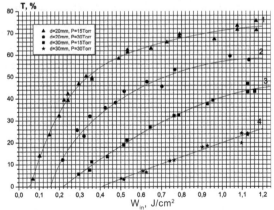

Fig. 3.17. Gas cell transmission dependence: 1 – cell 20 mm thick, 15 Torr; 2 – 20 mm thick cell, 30 Torr; 3 – cell 30 mm thick, 15 Torr; 4 – cell 30 mm thick, 30 Torr.

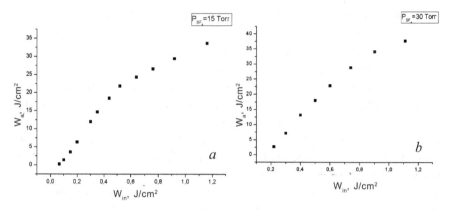

Fig. 3.18. Dependences of W_a energy absorbed in a 10 mm thick gas layer on the density of laser energy incident on a gas W_{in}: a) 15 Torr; b) 30 Torr.

10 mm thick gas layer on the density of the laser energy incident to the gas, which are shown in Fig. 3.18. Figure 3.18 presents dependences $W_a = f(W_{in})$, obtained at SF_6 pressures of 15 and 30 Torr. In special experiments, it was found that a small (up to 50% in pressure) addition to SF_6 of ethane and other gases (C_3F_8, He, N_2, Ne) has practically no effect on the absorption coefficient of CO_2 laser radiation. Therefore, the dependences presented in Fig. 3.18 were used further (see Section 3.3.4) to study the effect of absorbed laser energy on the SSVD characteristics in the SF_6 based mixtures.

3.2.3. Investigation of the effect of CO_2 laser radiation on the combustion voltage of SSVD in SF_6 and mixtures based on it

The effect of CO_2 laser radiation on the SSVD burning voltage was studied at various discharge current durations (τ_{dis} = 100÷500 ns) and the energy inputted from the pump capacitance into the SSVD plasma (W_{el} = 50÷150 J/l). Photographs of SSVD taken when the gap was not irradiated with CO_2 by a laser, and after irradiation with laser radiation, are shown in Fig. 3.19, and the oscillograms of discharge current and voltage pulses on the SSVD plasma and discharge current corresponding to these conditions are shown in Fig.3.20.

From photographs of the SSVD glow, it can be seen that after irradiating the discharge gap with laser radiation, the SSVD glow becomes weaker and filamentation of the discharge occurs. It is seen from Fig. 3.20 that when the discharge gap is irradiated with CO_2 laser radiation (W_{in} = 1.1 J/cm²), the voltage across the plasma

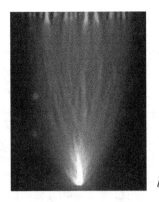

a *b*

Fig. 3.19. Photographs of the SSVD glow in an $SF_6:C_2H_6 = 5:1$ mixture with a total pressure of 18.2 Torr: a) the gap was not irradiated by CO_2 laser radiation; b) the discharge gap was pre-irradiated with laser radiation with $t = 4$ μs.

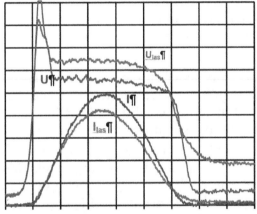

Fig. 3.20. Oscillograms of the voltage of U_{las} and the discharge current I_{las} after the gas is preliminarily irradiated with a laser $t = 4$ μs in an $SF_6:C_2H_6 = 5:1$ mixture at a total pressure of $P = 18$ Torr. *I, U* – oscillograms of current and voltage of non-irradiated discharge gap, respectively. Sweep 50 ns/div.

increases noticeably, and the discharge current decreases. The effect of CO_2 laser radiation on the combustion voltage of the SSVD was characterized by the magnitude of the relative increase in the combustion voltage $\Delta = (U_{las}-U_0)/U_0$, here U_{las} and U_0 are the combustion voltage of the SSVD when the gas is irradiated with CO_2 by the laser and without exposure, respectively. Most of the experiments were carried out at $\tau_{dis} = 200$ ns and $W_{el} = 50$ J/l, since an increase in τ_{dis} and W_{el} leads to a decrease in the stability of the SSVD and distortion of voltage oscillograms due to germination of the plasma channel from the cathode. To increase the stability of the

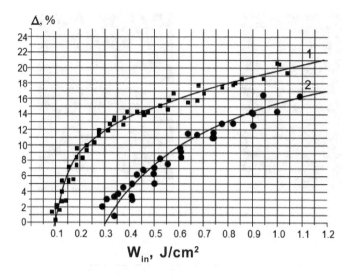

Fig. 3.21. Dependence of the relative increase in the discharge voltage of a discharge on the energy density of a CO_2 laser incident on a gas.

SSVD to SF_6, ethane was added in the ratio of partial pressures of SF_6:C_2H_6 = 5:1. The addition of ethane to SF_6 made it possible to obtain a stable SSVD under the condition of preliminary irradiation of the discharge gap by the radiation of a high-intensity CO_2 laser. In pure SF_6, already with the incident radiation density of W_{in} 0.8 J/cm², $\tau > 3$ µs and $\tau_{dis} > 200$ ns, the discharge broke into a spark, therefore, the dependences of $\Delta U = U_{las} - U_0$ and Δ on the density of the laser radiation were removed mainly for SF_6:C_2H_6 = 5:1 mixtures, which made it possible to investigate the effects of laser irradiation on the discharge characteristics in a wider range of parameters τ and W_{in}.

Figure 3.21 presented for the two mixtures (SF_6:C_2H_6 = 15:3 and SF_6:C_2H_6 = 30:3 Torr) shows the dependences of the relative increase in the combustion voltage Δ on the laser energy density incident on the gas (experimental points were obtained with the following parameters: $\tau = 3$ µs, $W_{el} = 50$ J/l, $\tau_{dis} = 200$ ns). This figure shows that with an increase in the density of the laser radiation incident on the discharge gap the discharge voltage increases. Since the VV and VT relaxation times in SF_6 are, respectively, $\tau_{VV} = 1.1$ µs Torr and $\tau_{VT} = 120$ µs Torr [129–131], we can assume that for the above values $P_{SF_6} \tau_{las}$ and τ almost all the energy of laser radiation absorbed by the gas goes into vibrational degrees of freedom. Then it can be assumed that the increase in the combustion voltage is due to the process of electron attachment to vibrationally excited SF_6 molecules, the rate of which is much greater than the rate of electron attachment to the SF_6 molecule, which is in the ground state. The increase in the density of excited SF_6 molecules is determined by the energy of the laser radiation absorbed in the discharge volume.

Figure 3.22 shows the dependences of Δ on the laser energy density absorbed in the discharge volume for two values of the partial pressure of SF_6. It can be seen from the above graph that an increase in the CO_2 laser energy absorbed in the discharge zone leads to an increase in the discharge burning voltage, while Δ for mixtures with a lower partial pressure of SF_6 is higher (with the same value of absorbed energy W_a). This is due to the fact that the proportion of absorbed laser energy per molecule is greater in this case. Figure 3.23 shows the dependences of Δ on the average number of photons of a CO_2 laser n_{ph} absorbed by one SF_6 molecule (for mixtures with an SF_6 partial pressure of 15 and 30 Torr). From Fig. 3.23 it can be seen that the dependences of Δ for the mixtures with different partial pressures of SF_6 have an identical form and quite well coincide with a low specific density of laser energy absorbed in the discharge volume. The difference manifests itself only at the densities of absorbed energy corresponding to the value of $n_{ph} > 7$. For $n_{ph} > 7$,

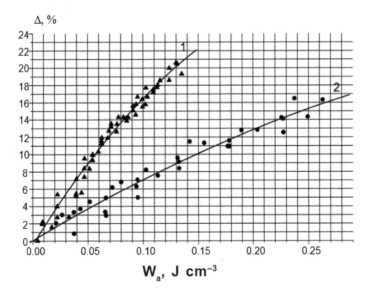

Fig. 3.22. Dependences of Δ on the laser energy density absorbed in the discharge volume: 1 – 15 Torr; 2 – 30 Torr SF_6.

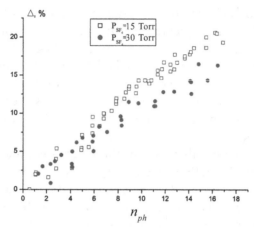

Fig. 3.23. Dependences of Δ on the average number of CO_2 laser quanta absorbed by the SF_6 molecule.

the value of Δ in mixtures with a large partial pressure of SF_6 is somewhat lower, which is apparently due to the processes of rapid VT relaxation of vibrational energy from highly excited levels of the SF_6 molecule, which, in turn, are populated in the VV exchange processes.

Indeed, the relaxation rate from high vibrational levels significantly exceeds the relaxation rate VT from the first vibrational level of the SF_6 molecule; therefore, even with $\tau \sim 3$ µs, the relaxation time VT for mixtures with high pressure will be noticeably less.

It should be noted that a decrease in Δ occurs when any gases are added to SF_6. Figure 3.24 shows the dependences of Δ obtained under the same experimental conditions ($P_{SF_6} = 6$ Torr, $W_{in} = 1.5$ J/ cm², $\tau_{dis} = 200$ ns, $W_{el} \approx 50$ J/cm³ and $\tau = 3$ µs) in SF_6 mixtures with different gases from the partial pressure of these additives.

From Fig.3.24 it is seen that, as in mixtures of SF_6 with He, the addition of molecular gases (C_2H_6 and C_3F_8) reduces the value of Δ. It is noteworthy that the molecular gases, both electropositive (C_2H_6) and electronegative (C_3F_8), reduce the value of Δ significantly more than He. Apparently, this is due to the more efficient deactivation of vibrationally excited states of SF_6 molecules by molecular gases in VV exchange processes, the rate of which is high under these experimental conditions.

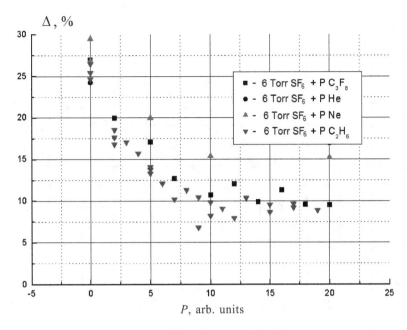

Fig. 3.24. Dependences of Δ in SF_6 mixtures with different gases on the partial pressure of this gas.

3.2.4. The role of the electron attachment process to vibrationally excited SF_6 molecules in the effect of limiting the current density

Based on the obtained data on the combustion voltage in a vibrationally excited SF_6 and the data in Section 3.1 of this book, we estimate the contribution that the attachment of electrons to the vibrationally excited SF_6 molecules makes to the effect of limiting the current density.

Figure 3.25 shows for the $SF_6:C_2H_6 = 15:3$ Torr mixture, the dependence of the relative increase in the combustion voltage $\Delta_{el} = (U_w - U_{50})/U_{50}$ on the specific energy input to the SSVD plasma W_{el}. Here, U_w is the discharge burning voltage at the specific energy input to the plasma SSVD $W_{el} = W$, U_{50} is the burning voltage of the SSVD at $W_{el} = 50$ J/l. It is clear from that figure that, just as in Fig. 3.22, with an increase in the energy input to the discharge volume, the value of ΔU increases, only in this case the energy density deposited into the discharge from the storage capacitance of the pump source and not the laser density energy absorbed in the discharge volume.

94

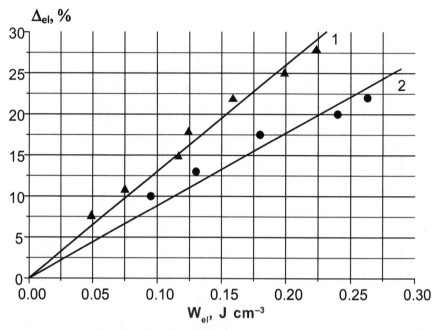

Fig. 3.25. Dependence of the relative increase in the burning voltage of the SSVD discharge in an SF_6:C_2H_6 = 15:3 mixture on the density of electrical energy absorbed in the discharge volume. 1 – mixture contains 15 Torr SF_6; 2 – mixture contains 30 Torr SF_6.

As noted earlier, under independent discharge conditions, the excitation of the vibrational degrees of freedom of the SF_6 molecules is less than 2% of the total energy deposited in the discharge [115, 116]. If we now set the value of the specific energy input W_{el} = 0.3 J/cm³, then only 0.006 J/cm³ goes to the excitation of the vibrational modes of the SF_6 molecules. Let us determine, using the graph presented in Fig. 3.22, what effect corresponds to this value, we see that Δ is only ~1% (that is, it lies at the limit of the registration of the 'effect'). In experiments to study the effect of limiting the current density, at a given energy input, the value of Δ_{el} is more than 15% (Fig. 3.8) Thus, it is necessary to recognize that in a normal SSVD, the process of attachment of electrons to the vibrationally excited SF_6 molecules is not the main one, its role in the growth of the burning voltage of the SSVD is insignificant.

3.2.5. Analysis of the results

As a result of experimental studies, it has been established that during irradiation of the discharge gap by CO_2 laser radiation, the combustion voltage of the SSVD in SF_6 and mixtures based on it increases due to the process of electron sticking to vibrationally excited SF_6 molecules. It has been established that with significant vibrational excitation of SF_6 molecules, the stability of SSVD decreases.

The dependences of the relative increase in the combustion voltage on the laser energy absorbed in the discharge volume are obtained. From these dependences, it follows that the burning voltage monotonically increases with increasing laser energy introduced into the discharge volume.

It has been established that an increase in the delay time between the irradiation of the working volume and the supply of a high-voltage pulse to the gap, as well as the addition of He, which accelerates the $V-T$ relaxation process, leads to a decrease in the effect of increasing the combustion voltage. In addition, the reduction of the relative increase in the combustion voltage results in the addition of any gases, which introduce an additional channel of relaxation of the vibrational energy stored by SF_6 molecules.

It is shown that when vibrating degrees of freedom of an SF_6 molecule is excited using a pulsed CO_2 laser, the growth of the voltage of the SSVD is significantly less than that observed when the same degree of vibrational excitation is achieved by increasing the current density of the SSVD. Therefore, under the conditions of a conventional self-sustaining discharge, the attachment of the electrons to the vibrationally excited SF_6 molecules is not the main process determining the increase in the discharge burning voltage with an increase in the specific energy input to the SSVD plasma. There are other processes whose contribution to the effect of increasing the combustion voltage of the SSVD is significantly higher than the attachment of electrons to vibrationally excited SF_6 molecules.

Conclusions to Chapter 3

The results of a study of the characteristics of a single diffuse channel in SF_6-based mixtures, obtained in Chapter 3, can be summarized as follows:

1. In SF_6 and mixtures based on it, the voltage across the plasma increases with increasing specific energy input, which directly indicates the existence of current density limiting mechanisms that prevent the passage of all the energy stored in the pump source into one diffuse channel.

2. Methods for studying the electrical characteristics of a single plasma formation (diffuse channel) have been proposed and implemented, which allow one to quantify the effect of limiting the current density and assess the contribution of the above-listed plasma-chemical processes to this effect. It has been established that the voltage on the plasma of a volume discharge in SF_6-based gas mixtures increases with an increase in the specific energy input, while the relative increase in the combustion voltage in SF_6 mixtures with hydrocarbons is significantly higher than in SF_6 mixtures with H_2. It is shown that using the criterion introduced according to this method, it is possible to qualitatively and quantitatively evaluate the effect of various gas additives on the stability of a volumetric self-discharge in SF_6-based gas mixtures.

3. It has been established that the main mechanisms for limiting the current density SIVD are: dissociation of SF_6 and other components of the working mixture by electron impact, leading to a decrease in the ionization rate and an increase in the electron attachment rate as a result of a local decrease in the E/N parameter in the diffuse channel and dissociative electron–ion recombination, leading to an increase in the rate of electron loss with increasing current density in the channel.

4. For the first time, SSVD was studied in SF_6 and mixtures based on it upon excitation of vibrational states of electronegative molecules by a pulsed CO_2 laser. A significant increase in the discharge burning voltage was found with an increase in the laser radiation energy absorbed by strongly electronegative gas molecules. It is shown that this effect is due to an increase in the rate of electron loss in the discharge plasma due to their adhesion to the vibrationally excited molecules.

5. It was found that the attachment of electrons to SF_6 molecules excited into vibrational states by electron impact in a plasma of a volume discharge does not significantly affect the balance of charged particles in a gas-discharge plasma. Therefore, electron attachment to vibrationally excited SF_6 cannot serve as a mechanism determining the development of an independent discharge in strongly electronegative gases in the form of SIVD.

4

Mechanisms of development of plasma instabilities of self-initiated volume discharge in working mixtures of non-chain HF(DF) lasers

In the fourth chapter, the effect of pulsed laser heating of gas on the characteristics of SIVD in SF_6-based gas mixtures is examined and the process of the development of detachment instability in active media of electric-discharge non-chain HF(DF) lasers due to electron impact from negative ions is investigated. The results presented in this chapter were obtained in [128, 129, 132–135, 140, 154].

4.1. Description of the experimental setup and experiment methodology

The effect of gas heating on the characteristics of SIVD in working mixtures of HF(DF) lasers, as well as in pure SF_6 and mixtures based on, was studied on the installation described in detail above (in section 3.2). The installation diagram is shown in Fig. 4.1. A spatially uniform beam of a CO_2 laser with dimensions of 60×60 mm was introduced into the discharge chamber through a window of BaF_2, part of the radiation was branched by NaCl wedges onto a calorimeter and a photodetector to control the energy and shape of the laser radiation pulse. The laser radiation energy was changed by

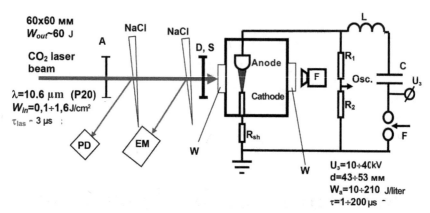

Fig. 4.1. Scheme of experimental setup A – attenuation films; NaCl – wedges from NaCl, branching part of the radiation into a calorimeter (CM) and a photodetector (PD); D, S — diaphragm or screen defining the profile of the discharge gap irradiation; W – BaF$_2$ windows; DC – digital camera; F – controlled discharger; C is the capacitance; L is the inductance; R$_1$, R$_2$ are resistive dividers; R$_{sh}$ – current shunt; Osc – digital oscilloscope.

calibrated teflon-based filters in stalled in front of the NaCl wedges. On the way of the laser beam, various apertures (screens) were installed, with the help of which areas with a given illumination profile were formed, or a diaphragm was used in the form of a slit, the areas adjacent to the electrodes were closed from the irradiation.

Figure 4.2 shows a photograph of the discharge gap through the entrance window a personal computer. The SIVD ignited between the needle cathode (lower electrode) and the lateral surface of the cylinder 15 mm diameter (anode) with an interelectrode distance of $d = 43$ mm (the upper electrode in Fig. 4.2). The role of the needle was performed by cutting a copper wire of 1.5 mm diameter in polyethylene insulation, which prevents the development of a discharge from the lateral surface of the cathode. In some experiments, when studying the structure of the SIVD in the irradiation zone, a tube with an outer diameter of 2 cm was attached to the cathode, which covered the cathode region in order to reduce the strong illumination from the channel developing from the cathode spot, as shown in Fig. 4.2*b*.

A capacitor with a capacitance of $C = 0.5 \div 2.6$ nF was discharged through the inductance $0.5 \div 4$ µH for the gap. The charging voltage varied from 15 to 35 kV depending on the pressure and composition of the gas mixture in the chamber. The moment of time when the voltage pulse was applied to the discharge chamber relative to the

Fig. 4.2. Photos of the discharge gap, taken through the input window of the discharge chamber.

laser pulse was regulated within fairly wide limits $\tau = 1\div200$ μs. The parameters of the CO_2 laser and the experimental conditions are described in Section 3.3 of this dissertation. The laser worked on the P20 line of a 10.6 μm band, the radiation pulse had a standard shape for CO_2 transverse discharge lasers, its full duration was $\eta_{las} \approx 3$ μs (Fig. 3.22). The energy density of the laser radiation absorbed by the SF_6-based gas mixture in the discharge development zone was $W_a = 0.1 \div 0.22$ J·cm^{-3}. The method for determining the absorbed energy W_a in the discharge region is described in detail in the section 3.3.3. The partial pressure of SF_6 in the mixtures studied was 9–15 Torr. The gas temperature T_g, which is established at the time of application of voltage to the discharge gap (after the end of the laser pulse), was determined from the expression

$$\frac{W_a}{N} = \int_{T_0}^{T_g} C_V(T')dT' \, , \quad \tilde{N}_V(T') = \sum_i \xi_i C_{Vi}(T').$$ (4.1)

where N is the total concentration of molecules, C_{Vi} is the heat capacity at constant volume, and ξ_i is the initial relative concentration of the i-th component of the mixture. Temperature $T_0 \approx 300$ K. The possibility of using this expression to determine T_g under the conditions of the experiments described below was theoretically and experimentally substantiated in Refs. [129, 132].

With the help of a digital camera, the opening of the shutter of which was synchronized with the supply of voltage to the discharge gap, the change in the structure of the SIVD during the preliminary irradiation of the discharge gap by a pulsed CO_2 laser was investigated. At each value of τ, as well as in the absence of an impact on the laser pulse gap, oscillograms of voltage and current of SIVD were taken. In the course of the experiments, the voltage was measured in the quasi-stationary phase of combustion of a self-sustained volume discharge U_{qs}, which ignited in SF_6-based mixtures previously heated by a pulsed TEA CO_2 laser. The change in the combustion voltage of the SIVD during gas irradiation with a laser was characterized by the parameter $\Delta = (U_{las}-U_{qs})/U_{qs}$, where U_{las} and U_{qs} are the values of the combustion stresses, respectively, in the irradiated and unirradiated discharge gap. For definiteness, U_{las} and U_{qs} were measured at time points corresponding to the maxima of the corresponding currents. The voltage and current of SIVD were controlled by a calibrated high-voltage voltage divider and a current shunt, respectively. Signals were recorded using a Tektronix 2012B and a TDS-220 oscilloscopes with a bandwidth of 100 MHz.

4.2 Characteristics of SIVD in SF_6 and mixtures based on it when exposing the discharge gap to the radiation of a pulsed CO_2 laser

4.2.1 Effects of CO_2 laser radiation on the spatial structure and stability of SIVD in SF_6 and mixtures based on it

Figure 4.3 shows diaphragms $D_1 \div D_6$ schematically mounted on the path of the laser beam (perpendicular to the optical axis). In this figure, the overlapped zones of the beam are shaded, the y axis (vertical) is directed from the cathode to the anode, the dimensions are indicated in mm. The results of the influence of laser irradiation on the structure of the discharge are presented in Fig. 4.4.

The discharge gap was photographed at fixed values $\tau = 3$ μs (the delay time of the application of a high-voltage pulse relative to the CO_2 laser pulse), $W_{in} = 0.8$ J/cm², U = 35 kV in an SF_6:C_2H_6:Ne = 15:6:15 Torr. Additives of C_2H_6 and Ne to SF_6 did not significantly affect the effects associated with the irradiation of the gap, but they allowed to increase the stability of SIVD (C_2H_6) and increase the intensity of the glow of the discharge (Ne).

As can be seen from Fig.4.4, the overlapping of a part of the beam with a screen leads to the complete displacement of the SIVD to the non-irradiated zone. In the irradiated zone, the luminescence of SIVD is not registered. When a circular diaphragm with a diameter

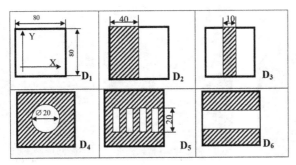

Fig. 4.3. Diaphragms mounted in the path of the laser beam.

smaller than the transverse size of the discharge (Fig.4.4g) is placed in the path of the beam, a 'hole' with a rather sharp edge appears in the photograph of the SIVD glow, and the discharge tends to go around the area illuminated by the laser. As can be seen from the comparison of Fig. 4.4*b* with Fig. 4.4*a*, under the influence of laser radiation, spatial forms of SIVD are observed, quite unusual for the geometry of the studied gap.

Since in the irradiation zone there are large electron losses due to sticking to vibrationally excited SF_6 molecules, the discharge does not burn in these zones. This allows one to control the structure of the discharge. If we place a diaphragm in the form of a grid in the laser beam (Fig. 4.3*d*), then the discharge will burn as a grid only in the shadow region (zones not exposed to radiation) (Fig.4.4*d*). If the diaphragm is made in the form of a slit whose size horizontally (X axis) is much larger than the transverse size of the discharge (Fig. 4.3*e*), SIVD can no longer bend around the irradiated zone, which is a kind of barrier in the gap

In this case, the brightness of the illumination in the irradiated zone is noticeably lower than the brightness of the rest of the discharge, despite the continuity of the SIVD current. It can be assumed that this circumstance is associated with an increase in the fraction of the ionic component in the plasma conductivity in the irradiation region due to the attachment of electrons to vibrationally excited SF_6 molecules (section 3.3.). Also noteworthy is the noticeable stratification of the SIVD in the irradiated zone in Fig.4.4*e*, which obviously should lead to a decrease in the stability of the SIVD. Indeed, special experiments confirm that pre-irradiation of the discharge gap by CO_2 laser radiation greatly reduces the stability of the SIVD in the SF_6-based mixtures. This fact is particularly evident in the study of the structure of the discharge,

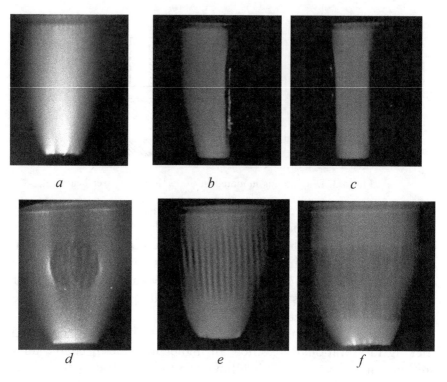

Fig. 4.4. Photos of SIVD.

when slotted diaphragms were installed in the beam (Fig. 4.3*d*). Visual observations of SIVD under these conditions show that in the area exposed to irradiation, the intensity of the discharge glow is weaker. The structure of the discharge also changes: there is a filamentation of a volume discharge, and the resulting structures depend on the density of the laser radiation incident on the gas. Figure 4.5 shows photographs of SIVD obtained in an $SF_6:C_2H_6:Ne = 15:3:15$ mixture at different values of the laser energy density absorbed in the irradiated discharge zone ($\tau = 4$ μs). In these photos, it is clearly seen that in the zone where the gas is heated by laser radiation, the discharge has a clearly distinguishable quasi-periodic channel structure, while reducing the intensity of irradiation (gas temperature) leads to an increase in the period of this structure. It is important to note that a significant increase in the density of the laser radiation incident on a gas (an increase in the gas temperature in the zone under consideration) leads to the development of plasma instability in the irradiation zone. We note here that at a radiation density of $W_{in} \geq 0.8$ J/cm², the formation of plasma channels is

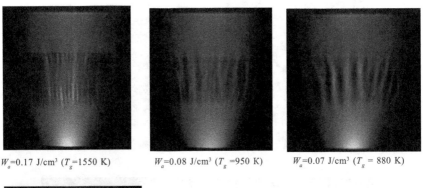

W_a=0.17 J/cm³ (T_g=1550 K) W_a=0.08 J/cm³ (T_g =950 K) W_a=0.07 J/cm³ (T_g = 880 K)

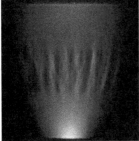

W_a=0.06 J/cm³ (T_g =800 K)

Fig. 4.5 Photos of SIVD illustrating the formation of quasi-periodic plasma structures in the gap, mixture: 15 Torr SF_6 +15 Torr Ne +3 Torr C_2H_6. The period of these structures depends on the intensity of irradiation, under each photograph there is the gas temperature in the irradiation zone and the energy density of the CO_2 laser absorbed in this layer is also indicated.

sometimes observed in the volume of the discharge gap (Fig. 4.6). The threshold value of the irradiation density at which the SIVD instability develops in the discharge gap volume depends on the duration and density of the SIVD current. Figure 4.6 shows a photograph of the SIVD in the discharge gap subjected to CO_2-laser irradiation, in which a bright plasma channel is seen. Here, unlike Fig. 4.5*a*, the density of the laser radiation absorbed in the discharge gap was 0.2 J/cm³ (otherwise, the experimental conditions were identical).

The same structure of the SIVD (with a plasma channel in the discharge gap volume) can also be obtained at lower gas temperatures in the heated region (T_g <1000 K) if the current density in the discharge gap increases, for example, by increasing the charging voltage or by shifting the region irradiation to the cathode. Due to the geometry of the needle–cylinder discharge gap, the current density near the cathode is significantly higher than that of the anode or in the middle of the discharge gap. Figure 4.7a presents a photograph illustrating the formation of plasma instability in the volume of the discharge gap with increasing current density in the irradiated discharge gap zone. The probability of occurrence of a spark channel in the heated region of the discharge gap increases

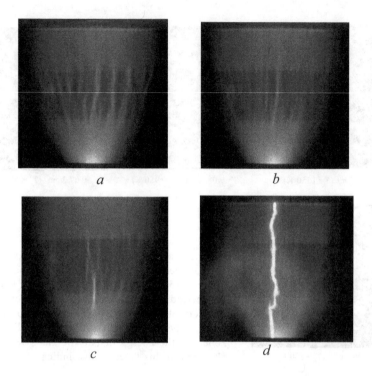

Fig. 4.6. Photos of SIVD illustrating the formation of quasi-periodic plasma structures and contraction of SIVD: *a*) T_g = 950 K; *b*) T_g = 1090 K; *c*) T_g = 1365 K; *d*) T_g = 1550 K.

significantly with increasing duration of the discharge current (in experiments, this was achieved by varying the parameters L and C of the discharge circuit).

It is noteworthy that without irradiation of the discharge gap by the laser, even with an increase in the energy input of W_{el} by 3 times (at this discharge duration, τ_{dis} = 180 ns), the SIVD remains fairly homogeneous, and similar plasma channels are not formed (SIVD looks like in Fig.4.4*a*) . (Details of the instability mechanisms of the SIVD, depicted in Figs. 4.6–4.7, will be discussed in section 4.4).

Figure 4.7*c* shows a photo of the SIVD (Fig.4.7*b*), when in the laser beam there was a diaphragm in the form of two slits 10 mm wide (Fig. 4.7*b*). This figure shows that a plasma channel was formed in the gap area closest to the cathode, but it didn't propagate into the gap area, closed off from the radiation by the diaphragm, and the channel did not form in the zone of the second gap (remote from the cathode). Note that in the discharge gap geometry used in the experiment, the current density is maximal near the cathode and

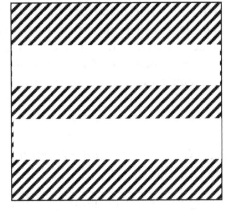

Fig. 4.7. *a*) Photograph of the SIVD in the mixture: SF_6:Ne:C_2H_6 = 15:15:3 Torr, the energy density of the CO_2 laser absorbed in the irradiation region of $W_a \sim 0.1$ J/cm^3. *b*) Image of the diaphragm overlapping the CO_2 laser beam. *c*) photograph of SIVD, obtained by irradiating the discharge gap through this aperture.

minimal at the anode. Therefore, from Fig. 4.7 we see that the plasma channel is formed when certain threshold conditions are fulfilled, both in terms of the gas temperature and the current density. Using a diaphragm in the form of a slit parallel to the surface of the anode, it is possible to determine the thresholds of instability development - by moving the diaphragm along the Y axis and varying the density of CO_2 laser radiation. Additional data on the development mechanisms of discharge instabilities in SF_6 can be obtained from experiments investigating the role of various gas additives.

As a result of studying the effect of various gas additives on the sustainability of SSVD (SIVD) in SF_6, it was found that when the CO2 laser was used to irradiate the discharge gap, the mixtures with a partial He content of more than 100% (relative to SF_6) have a markedly lower stability than the mixtures without He. This is probably due to the fact that He does not accelerate the processes of V–T relaxation in SF_6. An increase in the temperature of the mixture can, in turn, increase the contribution of such a process as the detachment of an electron from a negative ion, whose concentration in the plasma of highly electronegative gases significantly exceeds the electron concentration (see section 4.5). So unlike He, Ne, whose role in the processes of V–T relaxation is small, does not have such a noticeable effect on the stability of SIVD. The mechanism for developing the SIVD instability (SSVD) in SF_6-based mixtures will be discussed in more detail in Section 4.5.

4.2.2. SIVD in SF_6 and SF_6 mixtures with C_2H_6 under conditions of development of shock-wave perturbations

Investigations of the mutual influence of the discharge and gas-dynamic structures in the discharge volume have recently become increasingly relevant, due to the need to increase the average power and repetition rate of pulses used by gas-discharge lasers [52], as well as the ability to actively control gas flows in order to reduce the aerodynamic resistance of various fast moving objects. This section presents the study of the characteristics of SIVD in SF_6 and SF_6 mixtures with other gases (C_2H_6, Ne, He, C_3F_8, N_2) under the conditions of development in the discharge volume of medium perturbations caused by heating of the gas by a pulsed CO_2 laser. If in section 4.2.1, the change in the structure of the discharge in irradiated discharge gaps was investigated, when effects due to the formation of gas-dynamic gas perturbations ($\tau < 5$ μs) were not yet manifest, in this section the studies were carried out at large τ, when shock-wave perturbations started to form, caused by pulsed gas heating in the discharge gap.

Figure 4.8 shows typical photographs of the emission of SIVD in an SF_6:He mixture: $C_2H_6 = 15:3:15$ Torr, obtained at various values of τ at a fixed value $W_a = 0.13$ J/cm³, which illustrate the change in the type of SIVD during the propagation of a shock wave in the direction from the centre of the discharge gap to the electrodes. The coordinates of the characteristic boundaries (visually

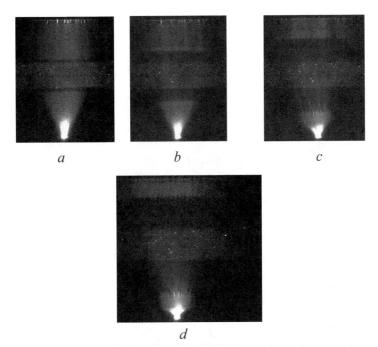

Fig. 4.8. Photos of SIVD at different values of the delay time τ between the voltage pulses on the discharge gap and the CO_2 laser: a) τ = 3.5 μs; b) τ = 20 μs; c) τ = 40 μs; d) τ = 50 μs. A mixture of SF_6:He:C_2H_6 = 15:3:15 Torr, irradiation through a slit diaphragm 10 mm wide; W_a = 0.215 J/cm³. X_1 and X_3 are the coordinates of the shock front, and X_2 and X_4 are the coordinates of the tangential discontinuity All X_1 coordinates were measured in relation to the cathode.

easily distinguishable) are also shown in Fig. 4.8: X_1 is the front of the shock wave propagating to the cathode; X_2 – the rear front of the shock wave propagating to the cathode; X_3 – the rear front of the shock wave propagating to the anode; X_4 – the front of the shock wave propagating to the anode. The dark band visible in the photographs of the glow of SIVD, which is located between the boundaries X_2 and X_3, corresponds to the heated gas, the density in this area decreases as the boundaries X_2 and X_3 move. The area between the boundaries X_1 and X_2 corresponds to a shock wave that moves towards the cathode; similarly, the zone between the borders X_3 and X_4 is a shock wave moving toward the anode. The density of the gas and the temperature in these areas are higher than the initial density and temperature of the cold gas. The process of establishing thermodynamic equilibrium, the distribution of characteristic thermodynamic parameters, and their temporal evolution in the

108

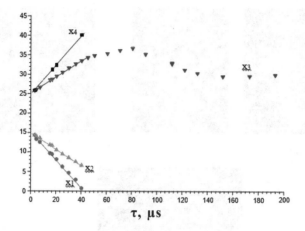

Fig. 4.9. The dependence of the coordinates of the shock-wave perturbation boundaries propagating in the discharge gap

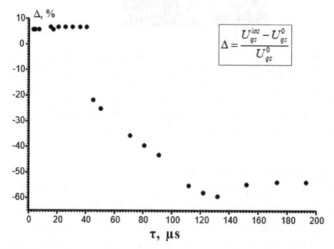

$$\Delta = \frac{U_{qs}^{las} - U_{qs}^{0}}{U_{qs}^{0}}$$

Fig. 4.10. Dependence of the parameter Δ on τ. A mixture of $SF_6:He:C_2H_6 = 15:15:3$ Torr at $W_{el} \approx 90$ J/l, $W_a = 184$ J/l.

zone of development of shock-wave perturbations were discussed in detail in [134].

Figures 4.9 and 4.10 show, respectively, the graphs of the dependences of the coordinates of characteristic boundaries X_i (τ) ($i = 1 \div 4$) of a shock-wave perturbation triggered by a pulsed CO_2 laser, and the dependence of the parameter on, taken in these same conditions. (The width of the irradiation region is H = 10 mm, the irradiation density is $W_{in} = 1.42$ J/cm^2). It is seen from Figs. 4.9 and

Non-irradiated gas   Irradiated gas   Non-irradiated gas

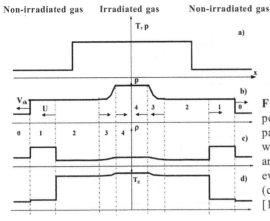

Fig. 4.11. The structure of the perturbation of thermodynamic parameters caused by the shock wave, the initial jump in temperature and pressure (a) and the subsequent evolution of pressure (b), density (c) and temperature (g) of gas [135].

Fig. 4.10 that the boundaries of the shock-wave perturbation move at a constant speed until the leading front of the perturbation reaches the electrode, followed by a spark breakdown of the layer of shock-compressed gas, and later using photographs it becomes difficult to identify the coordinates of the characteristic shock-wave boundaries. .

From Fig. 4.10, it can be seen that at the discharge gap subjected to irradiation, the combustion voltage of the SIVD is higher than in the non-irradiated gap during the whole time until the boundary of the shock-wave perturbation reaches the electrode. In [127–129], it was shown that an increase in the combustion voltage in an irradiated discharge gap is caused by an increase in electron losses in the processes of electron attachment to vibrationally excited SF_6 molecules. It should be noted that the dependences presented in Fig. 4.9 and Fig. 4.10 are typical for SF_6 and all gas mixtures based on it. With an increase in the density of the incident laser radiation and the width of the irradiation zone, the nature of these dependences does not qualitatively change, and the value of the parameter Δ linearly increases with an increase in the energy density W_a of the laser absorbed in the irradiated discharge gap zone. The velocities of displacement of the shock-wave perturbation boundaries are also uniquely determined by the composition of the gas mixture and the density of the absorbed energy W_a of the laser [132]. This circumstance makes it possible to determine the temperature of the gas in the irradiated zone with a good accuracy by the slope of the $x_i(\tau)$ dependences (in the linear section), established at the end of the laser pulse ($\tau_{las} \approx 3\mu s$) [134].

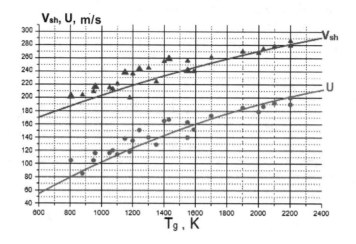

Fig. 4.12. Temperature dependences of V_{sh} and U. Points - experiment; lines are calculated by formulas (4.2) and (4.3). The gas temperature was determined according to the relation (4.1).

Figure 4.12 shows the dependences of the velocity of the leading front of the shock wave V_{sh} and the velocity of the trailing front U on the gas temperature in the irradiated region, the experimentally measured values are shown by points (determined on the basis of SIVD photos), and the lines show the results of calculations using the formulas [134, 135] :

$$\frac{T}{T_0}\left(1-\frac{\gamma-1}{2}\frac{U}{V_S}\right)^{\frac{2\gamma}{\gamma-1}}=\frac{2\gamma_0 M_0^2-(\gamma_0-1)}{\gamma_0+1}, \qquad (4.2)$$

$$U=M_0 V_{S0}\frac{2(M_0^2-1)}{M_0^2(\gamma_0+1)}. \qquad (4.3)$$

For the adiabatic exponent γ and the speed of sound V_S, subject to the constancy of the composition, the following simple relations are valid:

$$\gamma=1+R/C_V(T) \quad, \quad V_S=\sqrt{\gamma\frac{RT}{\mu}}, \quad \mu=\sum_i\xi_i\mu_i \qquad (4.4)$$

Here, U is the velocity of the contact surface between the regions of compression 1 and rarefaction 2, M_0 is the Mach number, μ_i is the molecular weight of the i-th component of the mixture, R is the universal gas constant. Since, as already noted, the gas density

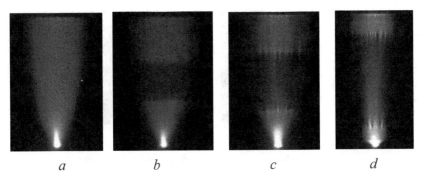

Fig. 4.13. Photos of SIVD in a mixture: SF_6:Ne = 15:3 with a total pressure of 18 Torr: *a*) – without laser irradiation; *b*) – τ = 4 µs; *c*) – τ = 20 µs; *d*) – τ = 50 µs. W_a = 170 J/l; T_g = 1800 K, V_{sh} ≈ 280 m/s.

in the region 4 $\rho = \rho_0$, the pressure drop initiating the shock wave $\Delta p = p-p_0$ is due exclusively to the temperature jump $\Delta T = T_g-T_0$. The latter, in turn, is completely determined by the value of W_a. Thus, from equations (4.5), (4.6) it follows that the shock wave velocities V_{Sh} and the interfaces between regions 1 and 2 U for given gas initial parameters are also functions solely of the value W_a. Because of this, they uniquely depend on W_a and all thermodynamic parameters of gas in regions 1–3 (Fig. 4.11). It is seen from Fig. 4.12 that the calculation results are in good agreement with the experiment – this is an additional confirmation of the validity of the previously made assumptions when calculating the gas temperature in accordance with relation (4.1).

Figure 4.13 shows photographs in the mixture: SF_6:Ne = 15:3 with a total pressure of 18 Torr, illustrating the ability to visualize shock-wave perturbations using a nanosecond discharge (τ_{dis} = 140 ns, W_{el}≈75 J/l). Here the discharge is forced to pass through 'barriers' – areas of shock-compressed gas. Figure 4.14*b-d* shows the photographs of the SIVD obtained in an SF_6:C_2H_6:He = 15:3:15 Torr mixture at different values of τ under conditions of laser beam limitation using the diaphragm shown in Fig.4.14*a*.

Figure 4.15 shows the discharge gap irradiation scheme and photographs of the SIVD obtained with different time delays relative to the laser pulse. The photos of Fig. 4.1 clearly show that at small times τ < 4 µs, the discharge is localized in the shadow region, and then, when a gas-dynamic perturbation is formed, the discharge begins to develop in the region where the rarefaction zone is formed

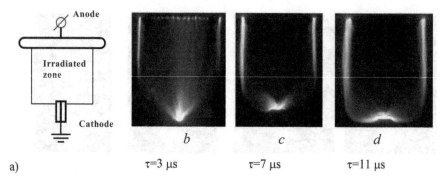

Fig. 4.14. a) schematic arrangement of the discharge gap and the irradiation zone; b) d) Photos of SIVD in the $SF_6:C_2H_6:He$ = 15:3:15 Torr mixture at different τ.

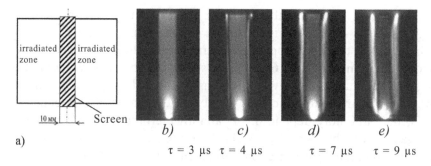

Fig. 4.15. a) Schematic representation of the irradiation area; b–e) - photographs of SIVD in an $SF_6:C_2H_6$ = 5:1 mixture with different τ. W_a = 0.11 J cm⁻³.

- in the photographs also, as in Fig. 4.13 characteristic 'whiskers' appear.

In general, the features of the structure of SIVD (SSVD), shown in the photos of Figs. 4.14 and 4.15, are trivially explained by the structure of the shock-wave perturbation formed by the diaphragm installed in the laser beam and the further evolution of this perturbation. The explanation of the dependence of the parameter Δ on τ, which is shown in Fig. 4.10, is not so obvious, and therefore requires some explanation. Here we present some qualitative considerations that clarify why the voltage in the gap remains constant until the shock-wave perturbation reaches the electrodes and what causes the appearance of the periodic structure of SIVD in the region with a higher gas temperature.

In [129], it was shown (see the next section of the book) that the dependence of $\Delta U = U_{las} - U_{qs}$ in SF_6 linearly depends on the gas temperature T_g in the irradiation region (measurements of U were carried out at $\tau = 4$ µs, while still in the discharge gap gas density is constant and not distorted by shock-wave perturbation). Thus, at constant pressure (concentration of molecules) of SF_6, the dependence $\Delta U (T_g)$ is approximated by the line:

$$\Delta U (T_g) = \text{const}(T_g - T_0), \qquad (4.5).$$

where T_g is the temperature to which the gas in the irradiation zone is heated using a CO_2 laser, and T_0 is the room temperature at which U_{qs} was measured. At constant T_g, the value of the parameter ΔU increases linearly with an increase in the width of the irradiation region H and the concentration of SF_6 (N).

It should be noted that (4.5) actually illustrates the increase in the pulse strength of the discharge gap due to an increase in the parameter $(E/N)_{cr}$ in SF_6-based mixtures when the gas is heated by a CO_2 laser [133]. Here E is the electric field, N is the concentration of gas molecules per unit volume. Since in this experiment (Fig. 4.10) N does not change, we have:

$$\Delta U = H \cdot N \cdot (E / N)_{cr}^{las} - H \cdot N \cdot (E / N)_{cr}^{0} = \Delta(E / N)_{cr} \cdot H \cdot N , \quad (4.6)$$

here $(E/N)_{cr}^{las}$ is the reduced electric critical field in the gas after laser heating, and $(E/N)_{cr}^{0}$ is the critical field in the non=irradiated gas. Thus, from the linear nature of the dependence $\Delta U(T_g)$ and (4.1) it follows that the dependence will also be linear:

$$\Delta(E / N)_{cr} = K_T \cdot \Delta T \qquad (4.7)$$

Here K_T is a constant, and $\Delta T = (T_g - T_0)$ is the temperature increase due to laser heating.

The time interval during which the voltage of the combustion of the SIVD on the irradiated gap exceeds U_{qs} in a cold gas, as follows from Fig. 4.10, is determined by the time of arrival of the shock wave to the electrode. After the front of the shock-wave perturbation crosses the electrode boundary, the voltage across the gap must begin to decrease, and this is due not only to the spark breakdown of the gas layer, but also due to a decrease in the density of gas molecules in the gap caused by gas-mass perturbation transfer. But until the shock-wave perturbation boundaries reach the electrodes, the value of U should remain almost unchanged, which follows from the following qualitative consideration.

The voltage across the gap, if we do not take into account the magnitude of the cathode potential drop is determined by the following expression:

$$U_{las} = \int_0^l (E/N)_{cr} (T_g, x) \cdot N \cdot dx \qquad (4.8)$$

$$\Delta U = \int_{x1}^{x4} \Delta(E/N)_{cr} (T_g, x) \cdot N(x) \cdot dx \qquad (4.9)$$

Given (4.3), we rewrite (4.5):

$$\Delta U = K_T \cdot \int^{x4} \Delta T(x) \cdot N(x) \cdot dx. \qquad (4.10)$$

Here x_1 and x_4 are the coordinates of the front of the shock wave propagating to the cathode and the anode, respectively. The expression under the integral in (4.10) is proportional to the amount of heat contained in the layer dx, in our case (radiative heat exchange can be neglected and $Q \approx W_{las}$), the integral in (4.10) is directly proportional to the laser energy W_{las} absorbed by the gas layer.

$$\Delta U = \text{const} \cdot W_{las} \qquad (4.11)$$

This is confirmed by the results of the experiments.

The formation of a quasi-periodic structure of the SIVD in the irradiated region (Figs. 4.8 and 4.13), apparently, is caused by the space charge formed at the boundary of the heated and cold gas zones. Indeed, the electric field E in the region of the heated gas is higher [129] than in a cold gas, therefore a space charge layer is necessarily formed on this boundary, whose density ρ is proportional to the value of ΔE:

$$\rho \approx (E/N)_{cr} N \sim W_a. \qquad (4.12)$$

Then for the period of structures l_c arising in the region of existence of a space charge we will have $l_c \sim \dfrac{1}{\sqrt{\rho}}$ or taking into account (4.12):

$$l_c \sim \frac{1}{\sqrt{W}} \qquad (4.13)$$

This agrees qualitatively with the results of the experiment shown in Fig. 4.5. The problem of the occurrence of periodic structures will be discussed more strictly in Section 4.4 of this book.

4.2.3. Analysis of results

Experimental data on the parameters of SIVD in SF_6 and its mixtures with other gases (C_2H_6, Ne, He, C_3F_8, N_2) under the conditions of formation in the discharge gap using a pulsed CO_2 laser of various medium perturbations (temperature, density) are presented. The change in the structure of a SIVD as a result of gas heating by laser radiation is investigated. It is shown that the laser irradiation of the discharge gap can lead to an increase in the combustion voltage of the SIVD (impulse strength of the gap) for some time until the leading edge caused by pulsed heating of the shock wave reaches the electrode boundary. The space-time evolution of a shock-wave perturbation propagating along the normal to the surface of the electrodes is investigated.

It is established that heating (a strong population of the vibrational states of SF_6) leads to discharge stratification, the appearance of filamentary structures in the irradiation zone and discharge contraction with an increase in the electrical energy introduced into the plasma.

4.3. Critical electric field strength in SF_6 and its mixtures with C_2H_6 at high temperatures

Of considerable interest is the question of the effect of gas temperature on the value of $(E/N)_{cr}$. This is connected with the solution of a number of fundamental and applied problems of modern electrophysics, such as the study of the dynamics of the high-temperature leader channel of a spark discharge and the use of highly electronegative gases as high-voltage insulation in conditions where operating temperatures can significantly exceed normal ones. SF_6 and mixtures based on it, which combine high dielectric strength with unique thermophysical properties in many respects, attract particular attention. However, information on the temperature dependence $(E/N)_{cr}$ in these gases up to 2006 was very limited and contradictory. There were only a small number of papers [13, 136–139] devoted to this issue. In [13], it is stated that the discharge ignition voltage in SF_6 at a constant gas density, and, consequently, the value $(E/N)_{cr}$ does not depend on temperature in the temperature range from room temperature to 1073 K. However, some experimental or theoretical evidence in favour of the statement in [13] is not given. On the contrary, experiments [136] on the passage of electrons in a uniform field through SF_6 gas heated up to a temperature of 600 K (direct measurements of the ignition voltage in [136] were not carried out) indicate a marked increase in $(E/N)_{cr}$ in SF_6 with increasing T_g. For example, according to [136], the value of $(E/N)_{cr}$ at $T_g = 600$ K should exceed the same value at room temperature by about 11%. The results of measurements of breakdown voltages in SF_6 at a pressure of 2 atm in the temperature range 1300–2200 K are presented in Refs. [137, 138]. They show a sharp drop in $(E/N)_{cr}$ with increasing temperature. A similar trend is clearly seen from the results of calculations [138, 139]. However, it should be noted that both these calculations and experiments [137, 138] correspond to the conditions when partial thermal dissociation of SF_6 takes place. Therefore, the values of $(E/N)_{cr}$, obtained in [137–139], do not refer to the actual gas, but to the mixture containing the products of its thermal decomposition. Measurements of $(E/N)_{cr}$ in SF_6 in the temperature range from 600 to 1300 K were not carried out before us. As for gas mixtures of SF_6 with C_2H_6, there was no information on the temperature dependence of the critical field strength in them before our work.

The experimental methods used in Refs [136–138] allow one to determine the reduced electric field strength in SF_6 heated to either relatively low temperatures [136] or to temperatures characteristic of an arc discharge [137,138]. Measuring $(E/N)_{cr}$ in SF_6 at intermediate temperatures requires a fundamentally different approach.

This section of the thesis is devoted to obtaining the temperature dependence of the reduced critical electric field strength in SF_6 and SF_6 mixtures with C_2H_6 heated by a pulsed CO_2 laser in the range $T_g = 293$–2400 K. Values $(E/N)_{cr}$ are estimated from the values of voltages in the quasistationary phase of combustion SSVD ignited in a gas with a certain time delay relative to the laser pulse. The gas temperature is determined from the measured specific values of the absorbed laser energy using the calculated specific heats of the mixture components. The temperature dependence of $(E/N)_{cr}$ in SF_6 is compared with the available literature data.

4.3.1 Experimental setup and measurement technique

The experimental setup and the experimental technique were similar to those described in [140] (see Section 4.4 of this book). During the experiments, the voltage was measured in the quasi-stationary phase of the combustion of the SIVD U_{qs}, which ignited in SF_6

and SF_6 mixtures with C_2H_6 at partial pressures of $SF_6 = 9$–30 mmHg, preliminarily irradiated by a CO_2-laser pulse. The energy density of the laser radiation absorbed by SF_6 in the zone of the discharge development reached $W_a = 0.27$ J cm^{-3}. The measurement technique of W_a was described in detail in Section 3.3 of this book. A spatially uniform beam of a CO_2 laser with dimensions of 60–60 mm was introduced into the discharge chamber through a window of BaF_2, part of the radiation was branched by NaCl wedges onto the calorimeter and the photodetector to control the energy and shape of the laser radiation pulse. The laser radiation energy was changed by teflon film filters installed in front of the NaCl wedges.

The SIVD ignited between the needle (cathode) and the lateral surface of the cylinder 15 mm in diameter with the interelectrode distance $d = 43$ mm. The role of the needle was performed by cutting a copper wire of 1.5 mm diameter in polyethylene insulation, which prevents the development of a discharge from the lateral surface of the cathode. The distance from the needle tip to the surface of the BaF_2 window along the optical axis inside the discharge chamber was 25 mm. A capacitor with a capacitance of 1 nF was discharged through the inductance of 2.5 µH at the discharge gap. The charging voltage varied from 15 to 35 kV depending on the pressure and composition of the gas mixture in the chamber. The delay between the laser and discharge pulses in these experiments was $\tau = 3$ µs, while the total duration of the laser pulse was also 3 µs. The delay was counted from the beginning of the leading edge of the laser pulse, which had the shape typical of a transverse-discharge CO_2 laser.

The magnitude of the quasi-stationary voltage U_{qs} was measured at the time instant corresponding to the maximum current. The specific contributions of the electric energy to the discharge plasma did not exceed 20 J/l, which guaranteed that the current density in the diffuse channel, which is characteristic of volumetric discharges in highly electronegative gases, does not affect the value of U_{qs} [120].

4.3.2 Analysis and discussion of the experimental results

Figure 4.4 shows the dependences of the voltage in the quasistationary phase of combustion of the SIVD U_{qs} on the specific absorbed energy W_a in the development zone of the discharge of the wave in SF_6 and the mixture SF_6:$C_2H_6 = 5$:1 at different values of the total pressure P. In the SIVD $(E/N)_{qs}$ phase, it is necessary to find the voltage

118

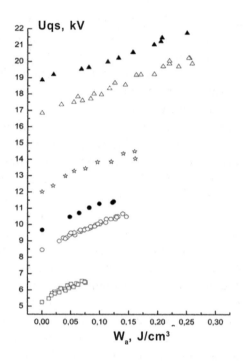

Fig. 4.16. Dependence of voltage in the quasi-stationary phase of combustion of the SIVD U_{qs} on the specific energy of the radiation absorbed in the development zone of the SIVD W_a: □ - discharge gap, P – 9 Torr; ○ – SF_6, P = 15 Torr; ▲ - SF_6 mixture: C_2H_6 = 5:1, P = 18 Torr; ★ – SF_6, P = 21 Torr; △ – SF_6, P = 30 Torr; ▲ – SF_6: C_2H_6 = 5:1 mixture, P = 36 Torr.

drop across the positive column of the SIVD. For this purpose, we have removed the dependences of U_{qs} on Pd at room temperature of the gases under study (in the absence of irradiation), from which it follows that $U_{qs} = \Delta U + \text{const} \cdot Pd$.

Neglecting the temperature dependence of ΔU, the quantity $(E/N)_{qs}$ was determined by the formula $(E/N)_{qs} = (U_{qs} - \Delta U) / (Nd)$. The dependences of $(E/N)_{qs}$ on the gas temperature T_g, constructed according to Figs. 4.16 and 4.17, are shown in Fig. 4.18. When calculating the T_g value in accordance with expression (4.1), the $C_V(T_g)$ values were taken from [141,142]. For comparison, Fig. 4.18 also shows the data [136] on the value of $(E/N)_{cr}$ in SF_6 at different temperatures. It can be seen that the results of measurements of $(E/N)_{cr}$ in [136] within the experimental error are in good agreement with our results for $(E/N)_{qs}$, however, our experimental method allowed us to significantly extend the temperature range compared

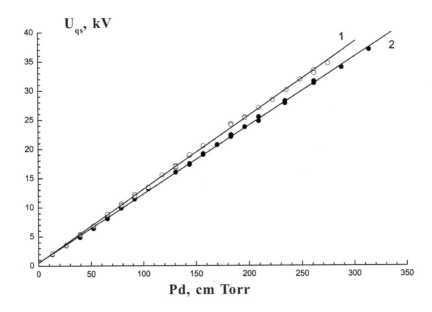

Fig. 4.17. Voltage dependences in the quasi-stationary phase of combustion of the SIVD U_{qs} on Pd: 1 - SF$_6$, 2 - SF$_6$:C$_2$H$_6$ = 5:1 mixture.

to [136]. The temperature dependences of $(E/N)_{qs}$ in the SF$_6$:C$_2$H$_6$ mixture, as follows from Fig. 4.18, have the same character as pure SF$_6$, i.e., the value of $(E/N)_{qs}$ increases with T_g in the temperature range studied.

For pure SF$_6$, the dependence of $(E/N)_{qs}$ on T_g was continued towards higher temperatures due to a small modification of the measurement technique for the $(E/N)_{qs}$ value [140]. The value $(E/N)_{cr}$ was determined on the basis of measurements of the combustion voltage of the SIVD ignited in the discharge gap, in a narrow zone of which, oriented perpendicular to the applied electric field, the SF$_6$ is preheated by the radiation of a pulsed CO$_2$ laser. The gas temperature was determined by two methods: by the energy of the laser radiation absorbed in the zone of development of the discharge, and by the velocity of the shock wave formed at the boundary of the heated and cold gas.

Thus, given that the values of $(E/N)_{qs}$ are close, as noted above, to $(E/N)_{cr}$, it can be argued that both our results and the data [136] clearly indicate an increase in the critical reduced strength of the electric field in SF$_6$ and SF$_6$ mixtures with C$_2$H$_6$ with increasing gas temperature, at least in the range T_g = 293–2400 K.

120

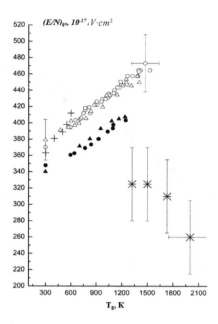

Fig. 4.18. Dependence of the magnitude $(E/N)_{qs} = (U_{qs}-\Delta U) / (Nd)$ on the gas temperature T_g: O – SF$_6$ P = 15 mm Hg; Δ – SF$_6$ P = 30 mm Hg; ▲ – SF$_6$:C$_2$H$_6$ = = 5:1, P = 18 mm Hg; ●– SF$_6$:C$_2$H$_6$ = 5:1, P = 36 mm Hg The critical value of the reduced field in SF$_6$ according to the works: + – [136], –✳ [138].

A completely different tendency is seen from the data on temperature dependence of $(E/N)_{cr}$, obtained in [137,138] and also presented in Fig. 4.18. It is seen that the value of $(E/N)_{cr}$ decreases rapidly with increasing T_g. As a result, even taking into account the rather significant measurement error obtained in [138], the value of $(E/N)_{cr}$, for example, at a temperature T_g = 1500 K, is noticeably lower than that found by us. The similar behaviour of $(E/N)_{cr}$ in SF$_6$ with increasing T_g also follows from the results of calculations given in [138,139]. They took into account the thermal dissociation of SF$_6$ molecules, the composition of the gas was considered to be thermally equilibrium, and the rate constants of ionization and adhesion were calculated using the energy distribution function of electrons. The latter was found by numerically solving the Boltzmann kinetic equation for electrons in an external electric field. According to [139], a decrease in $(E/N)_{cr}$ is associated with a relative increase in the concentration of fluorine atoms as the gas temperature rises. Since the elastic and inelastic energy losses of electrons during their collisions with fluorine atoms, including sticking processes, are

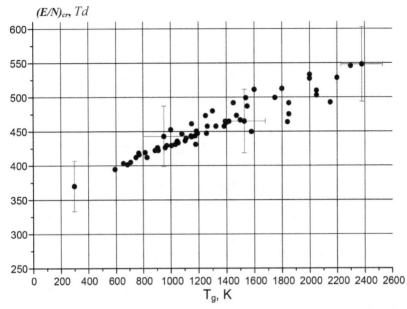

Fig, 4.19. Dependence of the magnitude $(E/N)_{cr}$ in SF_6 on temperature T $(P_{SF_6} = 15$–30 mmHg).

significantly less than when scattering electrons on SF_6 molecules and decomposition products (SF_4, SF_2, S_2, etc.), the ionization and sticking rates heated gas is achieved at lower values $(E/N)_{cr}$. compared to SF_6. This point of view is indirectly supported by the results of experiments [143], showing a decrease in $(E/N)_{cr}$ in SF_6 mixtures with He with an increase in the relative concentration of He. Indeed, the energy loss of electrons on helium atoms, as well as on fluorine atoms, is significantly less than on SF_6 molecules. The reason for the qualitative difference in the results of the work [137,138] from that obtained in the thesis consists, apparently, in the following.

In our experiments, the half-amplitude discharge current duration was ~100 ns, while the laser pulse duration and the time delay between the laser and high-voltage pulses were $\tau = 3$ μs. Therefore, the dissociation of SF_6 molecules could have a noticeable effect on the value of $(E/N)_{cr}$ only if it had time to occur on a time scale of ~3 μs. Dissociation of polyatomic molecules like SF_6 is a very complex process and usually involves many sequential decomposition stages of a polyatomic molecule. Unfortunately, there is currently no complete information on the SF_6 dissociation stages and the corresponding constants. One can, however, obtain a fairly reliable estimate of the dissociation time τ_d, based on the following qualitative considerations,

It is clear that the value of τ_d must be at least not less than the characteristic time τ_A of the single stage of the process. The corresponding constant k_a can be roughly estimated using the well-known Arrhenius relation [144]

$$k_A = S\left(\frac{8T_g}{\pi\mu}\right)^{1/2}\sigma_A\left(1+\frac{D_A}{T_g}\right)\exp\left(-\frac{D_A}{T_g}\right) \qquad (4.14)$$

Here S is the steric factor that takes into account the different mutual orientations of the colliding particles, μ and σ_A are their reduced mass and collision cross section. D_A is the activation energy corresponding to the equilibrium path of the reaction. The factor $S = 0.1–1$ for collisions of atoms with each other and with molecules and $S = 10^{-6}–10^{-3}$ for collisions of molecules with each other and with radicals. The value of D_A depends on which fragments a molecule breaks up in a single collisional act and for SF_6 it is of the order of several electron volts. This D_A estimate is in agreement with the data [139,145] on the equilibrium composition of heated SF_6 and the empirical relationship between the activation energy and temperature at the stage of developed dissociation [146]. Assuming for estimates $D_A \sim 10–15$ cm^2 and $S = 1$, we come to a strongly overestimated value $k_a \sim 10–14$ cm$^3\cdot$s^{-1}. Since the maximum particle concentration in our experiments does not exceed 10^{18} cm^{-3}, the value of D_A and, therefore, d is certainly greater than 100 μs. This allows us to assert with confidence that the dissociation of SF_6 molecules under the conditions of this experiment does not occur. Note also that in work [137] the dissociation of SF_6 became noticeable after its heating for 10 ms at a temperature of ~2000 K. This can be considered as an additional indirect confirmation of the above τ_{dis} estimate. In experiments [136], the dissociation of SF_6 was absent due to the extremely low gas density.

The increase in the critical value of the reduced field strength $(E/N)_{cr}$ in undissociated SF_6 with increasing gas temperature T_g observed in our experiments is associated with an increase in electron loss due to their attachment to the vibrationally excited SF_6 molecules. This follows from [128, 129]. At temperatures from 300 to 1100 K, dissociation in SF_6 is completely absent [13, 139]. Therefore, the critical field strength in SF_6 at a constant gas density always depends on its temperature in the specified range, namely, it

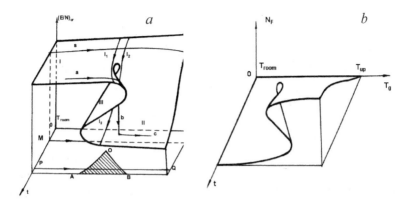

Fig. 4.20. Assembly catastrophe on temperature dependence $(E/N)_{cr}$ in SF_6: *a*) Assembly crash on temperature dependence $(E/N)_{cr}$ in SF_6. T_g is the gas temperature and *t* is the time during which the gas was at this temperature. The curves AO and BO are the set of bifurcation points of the catastrophe assembly, the arrow indicates the direction of movement of the points and their projections onto the plane of control parameters *b*) The dependence of the concentration of fluorine atoms NF on the gas temperature T_g and the heating time *t*.

increases with increasing T_g. This is in sharp contradiction with the statement of the author [13].

Thus, based on the analysis of literature data and experimental results of the present studies on the temperature dependence of the critical electric field in SF_6, it is shown that the dependence $(E/N)_{cr}$ in SF_6 on the plane of the parameters T_g and *t* (T_g is the temperature, *t* is the time from the moment of pulse heating of the gas) has a singularity, which is topologically represented by a feature of only one type – 'assembly catastrophe'. An interpretation of this phenomenon is given. The catastrophe on the $(E/N)_{cr}$ dependence in the plane of the T_g and *t* parameters is due to the fact that during pulsed heating, conditions are created under which an equilibrium is reached between the translational and internal degrees of freedom of the SF_6 molecule, but there is still no chemical equilibrium. As a result, the degree of dissociation of SF_6 is a function of not only of temperature, but also of time (from the moment of heating). In turn, the time to establish chemical equilibrium strongly depends on the temperature of the gas. The ambiguity of the dependence of $(E/N)_{cr}$ on T_g is due to the fact that the critical field in the SF_6 dissociation products is significantly less than in the non-dissociated gas, and in the heated but not dissociated SF_6 $(E/N)_{cr}$ is higher than at room temperature, due to the higher cross section for the attachment of electrons to vibrationally excited molecules. Figure 4.20 shows qualitatively the dependence of $(E/N)_{cr}$ on T_g and t.

In this work, temperature dependences of the critical value of the reduced electric field strength $(E/N)_{cr}$ in SF_6 and SF_6 mixtures with C_2H_6, previously heated by a pulsed CO_2-laser, were obtained. It is shown that the value of $(E/N)_{cr}$ in these gases increases with increasing temperature T_g in the range $T_g = 293$–2400 K. This agrees with the results of measurements of $(E/N)_{cr}$ in SF_6 performed in [136] at temperatures up to 600 K, and is in contradiction with the data [13] on the temperature dependence of $(E/N)_{cr}$ in SF_6 in the temperature range from room temperature to 1073 K. The growth of $(E/N)_{cr}$ with increasing gas temperature under the conditions of the present experiment is due to additional losses electrons as a result of their sticking to the vibrationally excited SF_6 molecules.

4.4. Negative instability in active media of electric-discharge non-chain HF(DF) lasers

The study of the ionization instability of a volume self-sustaining discharge in SF_6 and its mixtures is of interest in connection with the development of non-chain chemical HF(DF) lasers initiated by SSVD [3, 54].

Currently, several mechanisms of ionization instability in electronegative gases are known. A general theoretical approach to their study was formulated in [146]. The development of instabilities in working media of CO_2 lasers due to the detachment of electrons from negative ions by neutral and electronically excited gas components was considered in [147–149]. As applied to excimer lasers, the mechanism of instability due to the dissociation of a small impurity of the electronegative component by electron impact ('burning out' of a halogen additive) [150–153] has been studied in detail. In SF_6, according to [103], the instability should be attributed to the stepwise ionization process of the SF_6 molecules.

A qualitatively different scenario of the development of ionization instability, due to the separation of electrons from negative ions by electron impact, is possible in highly electronegative polyatomic gases at medium pressures and on a time scale of the order of several tens of nanoseconds. For the first time this question was considered in [154] as applied to SIVD in SF_6 and its mixtures. Once again, we note that in our conditions, the characteristics of SIVD and SSVD with pre-ionization are identical, but since the term SSVD is more general, then this section will use the term SSVD to emphasize the fact that it is possible to implement sticky instability, including in discharges with pre-ionization of the discharge gap. Of fundamental importance in this regard is the following circumstance. In gas mixtures of medium pressure based on SF_6 at room temperature, the best agreement between the calculated and experimentally recorded oscillograms of the current and voltage of SSVD (SIVD), including its quasistationary phase at $(E/N) \approx (E/N)_{cr}$, is observed, as will be shown below, when selected. Here and are the rate constants of electron-ion recombination and the separation of electrons from negative ions by electron impact, respectively, is the critical value of the reduced electric field strength. In this case, the increase in the electron density due to the destruction of negative ions by electron impact is almost completely compensated by their death during dissociative electron-ion recombination. The nonlinear electron multiplication mechanism mentioned above begins to manifest itself only with a noticeable imbalance of the constants and, which is possible either with a significant heating of the gas, or with $(E/N) \gg (E/N)_{cr}$. Indeed, the constant β_{ei} decreases in this case $(\beta_{ei} \sim T_g^{-1} T_e^{-\chi}$, $\chi > 0$, T_g is the gas temperature, $T_e = 2\langle\varepsilon\rangle/3$ is the average electron energy [155]), whereas the value k_d can only increase.

The conditions under which $(E/N) \gg (E/N)_{cr}$ are achieved, for example, near the tip of an incomplete channel, sprouting from the cathode. However, in this case, ionization processes develop on a spatial scale of the order of the channel radius, which, in an experimental study, leads to considerable difficulties. At the same time, it is possible to realize a noticeable imbalance of the constants and, thus, initiating the considered mechanism of otlipital instability in large volumes of gas. In this respect, the heating of gas mixtures based on SF_6 with a pulsed CO_2 laser is effective. We applied this approach in order to study the instability of the SSVD due to electron separation from negative ions by electron impact in SF_6 and gas mixtures based on it, including in the working media of HF(DF) lasers.

4.4.1. Experimental setup and experimental results

The experimental setup (Fig. 4.1) and the experimental procedure were the same as in section 4.1. Gas was pre-heated only in a narrow discharge gap zone. This was achieved by irradiating the discharge gap with a pulsed CO_2 laser through a slit diaphragm 10 mm wide, oriented perpendicular to the direction of the applied electric field. As will be seen from the material presented below, such an irradiation scheme makes it possible to observe the SSVD contraction directly in the volume of the heated gas (similarly to the glow discharge contraction [156]), and not as a channel sprouting from the cathode spot and as the spanning gap develops [44]. The gas temperature, which is established in the SSVD burning zone, was determined by the laser radiation energy absorbed by the SF_6 molecules [129], and also by the velocity of propagation of the shock wave formed at the boundaries of the cold and heated gas [135]. The temperature was varied in the range T_g = 800–2100 K (the specific energy of the laser radiation absorbed by SF_6 in the SSVD burning region, W_a = 0.05–0.23 J cm^{-3}). The voltage pulse on the discharge gap was applied with a delay of 4 μs relative to the laser pulse, which ensured the establishment of thermal equilibrium between the translational and internal degrees of freedom in the irradiated gas at the beginning of the discharge in the pressure range studied here [129]. In the course of the experiments, the voltage at the discharge gap and the current of the SSVD were controlled, respectively, and the SSVD was measured with a digital camera synchronized with a laser pulse. In order to identify the main processes that determine the current–voltage characteristics of the SSVD, oscillograms of voltage and current of a limited discharge were also taken. To this end, the SSVD was ignited in a quartz tube with a diameter of 8.5 mm with an interelectrode distance of 43 mm (see Chapter 3). The experimental oscillograms were compared with the calculation (the calculation method is described in detail in [100]).

Figure 4.21 shows photographs of the SSVD (Fig. 4.21 *a, b*) taken in an SF_6:Ne:C_2H_6 mixture at = 0.2 J cm^{-3}, τ_{dis} = 150 ns and different values, and the distribution of the emission intensity of the discharge plasma corresponding to the photographs along the X axis (Fig. 4.21c, d), parallel to the boundaries of the heating zone and passing through its middle. From Fig.4.21, it is clear that in the region selected by the heating, the SSVD stratifies, acquiring a current structure close to periodic. The spatial period of the structure formed decreases with increasing. Attention is drawn to the fact that the large current formations in Figure 4.21 *b*, in turn, consist of thinner filaments. Despite the separation of the SSVD, the plasma channels in the heating zone at not too high values and duration of the discharge current have a diffuse character. Increase of W_a or τ_{dis}

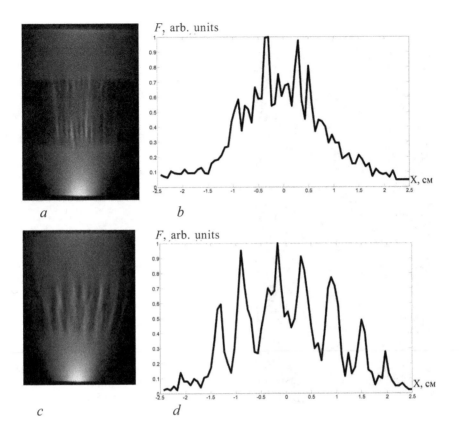

Fig. 4.21. a), b) photos of the SSVD at T_g = 1550 K (*a*), T_g = 800 K (*b*); c), d) -distribution of the intensity of the SSVD luminescence F along the coordinate in the heating zone at T_g = 1550 K (*c*) and T_g = 800 K (*d*). A mixture of SF_6:Ne:C_2H_6 = 5:5:1, p = 33 Torr, W_a = 0.2 J cm^{-3}, τ_{dis} = 150 ns.

leads to the release of one of the channels, usually located near the centre line of the electric field, and its contraction.

It is characteristic that, as in the case of cold gas, when heated, the SSVD contraction threshold in terms of parameters and in pure SF_6 is lower than in the SF_6:C_2H_6 mixtures. The above is illustrated in Fig. 4.22, where a photograph of an SSVD in pure SF_6 is shown, taken at τ = 160 ns and W_{el} = 0.12 J cm^{-3}. It can be seen from this figure that the instability develops directly in the volume of the discharge gap separated by laser heating of the gas. Apparently, an analogy with the contraction of a low-pressure glow discharge is appropriate here [155,156]. With a further increase of W_{el} or τ_{dis} the

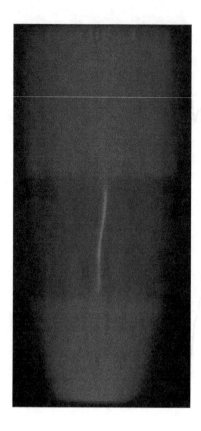

Fig. 4.22. A photograph of the SSVD in SF6 at = 0.12 J cm⁻³, T_g = 1150 K, τ_{dis} = 160 ns and p = 15 Torr.

spark channel closes the entire discharge gap. When W_{el} or τ_{dis} are constant the likelihood of contraction of the SSVD also increases with growth in the irradiated zone of the discharge gap.

Figure 4.23 *a* shows the oscillogram of the voltage of the limited SSVD in SF_6 at a pressure of p = 15 Torr and W_{el} = 0.12 J cm⁻³. Figure 4.23*b* presents the calculated oscillograms of voltage, upon receipt of which the model took into account the following processes: ionization by electron impact and electron attachment (IA); ion–ion recombination (rate constant β_a = 2 × 10⁻⁸ cm³ s⁻¹ [20]); dissociative electron–ion recombination (EIR) (the value in the calculation varied in the range = 0.5–3 × 10⁻⁷ cm³s⁻¹); separation of electrons from negative ions by electron impact (SE, k_d = 3 × 10⁻⁷ cm³ s⁻¹ [157]); SF_6 dissociation by electron impact (the price of the formation of a fluorine atom in the works of various authors is ~4–6 eV [50,51]). The calculated current in Fig.4.23 *b* is given for the case when all the listed processes are taken into account. The attachment of electrons to vibrationally excited SF_6 molecules was not considered, since we

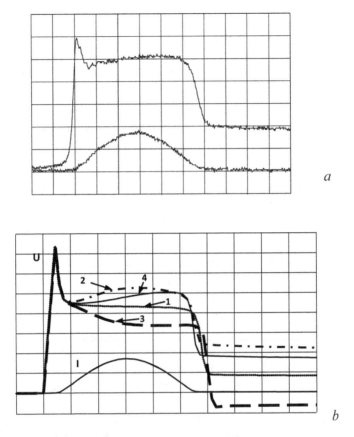

Fig. 4.23. Oscillograms of voltage and current of SSVD in SF$_6$ at p = 15 Torr and an interelectrode distance of 43 mm (limited discharge): a) – experimental oscillograms; b) – calculation (excluding cathode drop). Scale: U = 2 kV / div, I = 80A/div, scan 50 ns/div. The calculation took into account the processes: 1 – IA; 2 – IA, EIR; 3 – IA, DE; 4 – IA, DEU at $k_d = b_{ei}$.

had previously established experimentally that this process does not make a significant contribution to the overall balance of charged particles in the SSVD plasma in SF$_6$. Comparison of the calculated and experimental oscillograms of the voltage of the SSVD shows that their closest matching is achieved at $k_d \approx \beta_{ei}$.

Thus, it is possible to single out the three most important for the further analysis of the mechanisms of development of the SSVD instability in the gas mixtures based on SF$_6$, including in working mixtures of HF lasers, the result: 1) The SSVD in the heating zone is stratified to form quasi-periodic plasma structures the structures formed depend on; 2) the development of the SSVD instability begins with its contraction directly in the volume of the discharge gap, separated by laser heating of the gas; 3) an increase in the electron concentration due to their detachment from negative ions by electron impact at room temperature of the gas is compensated for by losses in the process of electron–ion recombination ($k_d \approx \beta_{ei}$).

4.4.2. Non-linear mechanism for the development of ionization in active media HF (DF) lasers

SF_6^- and SF_5^- ions predominant in the SSVD plasma are initially formed as a result of the attachment of electrons to SF_6 molecules [116].

The non-linear mechanism of electron generation in SF_6 due to the destruction of negative ions by electron impact was first considered in Ref. [157]. In the same place, the electron impact $k_d(SF_6^-) = 3 \times 10^{-7}$ cm^3 s^{-1} estimate for the electron detachment velocity of the negative SF_6 ion is obtained; The latter, as is known, exceeds 10^{-15} cm^2 [127]. It was also taken into account that at $(E/N) \sim (E/N)_{cr}$ the average electron energy $<\varepsilon> \sim 8$–10 eV is significantly higher than the electron binding energy in the negative SF_6^- ion (0.65–1eV [116]), as a result, the process of destruction of SF_6^- ions by electron impact can be considered thresholdless.

As for SF_5^- ions having a binding energy of $E_b \sim 2.8$ eV [116], the corresponding value $k_d(SF_5^-)$, taking into account the Boltzmann factor, $\exp(-E_b/T^*)$, $T^*2<\varepsilon>/3$ is less than $k_d(SF_6^-)$, but not more than 40%. In the quantitative description of the processes of separation of electrons from SF_6^- and SF_5^- by electron impact, this makes it possible to use the approximation of negative ions of the same type and introduce, respectively, a single rate constant of separation.

Of fundamental importance is the fact that no other process of detachment of electrons can compete with the destruction of negative ions by electron impact in the above range of pressures and times. Indeed, at values of heavy polyatomic negative ions, with the exception of their negligible share in the 'tail' of the energy spectrum, they do not have the energy required for electron detachment in collisions with neutral molecules [116], and the relatively low density of the gas virtually eliminates the effect of electronically excited components on the process of destruction of negative ions.

In [154], on the basis of exact analytical solutions of a nonlinear integro-differential equation, expressions for the electron density are obtained. The nature of the increase depends on the parameter [154]:

$$\xi = (a^2/\lambda - 2)/\lambda, a = (\alpha/\eta - 1), \ \lambda = n_e(0)(k_d - \beta_{ei})/(\eta u_e) \ (4.15)$$

$n_e(0)$ where is the electron concentration, established at the beginning of the quasistationary phase.

If $\xi > 0$, then

$$n_e(t) = n_e(0)(2b^2\lambda A)/(1-A)^2,$$

$$A = [(a/\lambda - b)/(a/\lambda + b)]\exp(b\lambda\eta u_e t) \ b = \sqrt{\xi}$$

$$(4.16)$$

In this case, the nonlinear multiplication of electrons occurs against the background of a noticeable contribution from the linear processes of impact ionization and sticking. It is no coincidence, therefore, that the solution (4.16) remains valid and for arbitrarily small values of $n_e(0)$ and/or $(k_d - \beta_{ei})$ and in the limit passes, as one would expect, into the well-known 'classical' expression for an electronegative gas [44, 45]:

$$n_e(t) \approx n_e(0)\exp((\alpha - \eta)u_e t) \tag{4.17}$$

At $\xi < 0$, the process of electron detachment from negative ions by electron impact plays a dominant role and the expression obtained for electron density

$$n_e(t) = n_e(0)\{\cos^2(b_1\lambda\eta u_e t)[1 - (a/b_1\lambda)tg(b_1\lambda\eta u_e t/2)]\}^{-1}, \ b_1 = \sqrt{-\xi}$$
$$(4.18)$$

for any values $n_e(0)$ $(kd - \beta_{ei})$ does not go to (4.17).

From relations (4.15), (4.16) and (4.18), it follows that with the imbalance of separation and recombination constants $((kd - \beta_{ei}) > 0)$, at the stage of increasing the discharge current $(\alpha > \eta)$, both solutions (4.16) and (4.18) have the pronounced 'explosive' character, i.e. some time after the start of the ionization process, the concentration of electrons formally becomes arbitrarily large. From (4.16) and (4.18), it is easy to find the corresponding characteristic 'explosion' times τ^i_{exp}. If $\xi > 0$, then

$$\tau^i_{exp} = \ln[(a/\lambda + b)/(a/\lambda - b)]/b\lambda\eta u_e \tag{4.19}$$

At $\xi < 0$

$$\tau_{exp}^i = 2arctg(\lambda b_1 / a) / (\lambda b_1 \eta u_e) \qquad (4.20)$$

In fact, of course, the result obtained only means that after a certain time $\tau_d^i \sim \tau_{exp}^i$ the process of electron attachment is partially compensated by their separation from negative ions, and therefore the electron multiplication rate increases dramatically. By analogy with the terminology used in the theory of excimer lasers, this could be called the 'burnout' of electronegativity, since in this case it is not about the destruction of electronegative molecules, but about the loss by gas, due to the process of detaching electrons, to be electronegative.

Although the derivation of relations (4.16) and (4.18) did not take into account the insignificant change in time of magnitude of $(\alpha - \eta)$, however, the 'explosive' nature of these decisions, in a certain sense, limiting to an exact dependence $n_e(t)$, indicates that this dependence has the same feature.

At the stage of discharge current decay in the quasi-stationary phase SSVD $\alpha < \eta$ and, respectively, $a < 0$. From relations (4.16) and (4.18) it follows that the concentration of electrons with time in this case always tends to zero. In other words, the process of volume multiplication of electrons as a result of their separation from negative ions by electron impact cannot compete with the loss of electrons due to attachment, if the capture of electrons by molecules is more effective than impact ionization. Thus, in the quasi-stationary phase of the SSVD, the ionization instability in the plasma volume, associated with the 'explosive' nature of the electron multiplication process, can develop only at the stage of increase of the discharge current.

4.4.3. Self-organization of SSVD plasma during laser heating of SF_6-based mixtures

The nature of self-organization in SSVD plasma in heated SF_6-based gas mixtures is currently not completely clear. There are, however, grounds for believing that this phenomenon can be significantly associated with the development of ionization instability in the SSVD plasma, which is caused by the electron impact electron-separation process discussed above. In this regard, we present some qualitative considerations.

At high gas temperatures, $(k_d - \beta_{ei}) > 0$, which at the stage of increasing the discharge current $(\alpha < \eta)$ leads to an 'explosive' increase. The time scale of the current change in the quasi-stationary phase of the SSVD is controlled by the LC chain $(\tau_c \sim \sqrt{LC})$ and under the conditions of this experiment is on the order of several hundred nanoseconds, then as $\tau_d^i \sim$ 20–30 ns. Since $\tau_c \gg \tau_d^i$, then in the quasi-stationary phase of the SSVD (when $(E/N) \approx (E/N)_{cr}$) this should inevitably lead to a redistribution of current in the volume of the discharge gap with the formation of current structures in the form of separate cords (for large values of T_g – thin filaments) Indeed, the standard linear analysis shows that it is the transverse-current spatially inhomogeneous perturbations that grow with the greatest increment under the conditions under consideration. At the nonlinear stage of development of perturbations, the formation of quasi-periodic structures observed in the experiment is possible. [158], it can be shown that such structures in the form of individual current cords prove to be stable. With increasing gas temperature, the difference $(k_d \sim \beta_{ei})$ also increases, which leads to a more rapid increase of $n_e(t)$. It can be assumed that it is for this reason that the number of conducting channels increases with increase of T_g and, accordingly, the spatial period of the current structure decreases. The experimental data presented in Fig. 4.21, in any case, do not contradict this assumption. An additional argument in favor of the above reasoning can be the fact that at extremely small values of T_g realized in the experiment, quasi-periodic current structures did not arise at all. Indeed, in this case, $\tau_{id} \gg \tau_c$.

At the stage of discharge current decay, the "explosive" mechanism discussed above does not work, since $(E/N) > (E/N)_{cr}$ and $\alpha < \eta$ (see Section 4.1) In this case, instability can develop in the falling portion of the quasistatic current–voltage characteristic $(I–V)$ of the discharge gap (time to establishment of $I–V$ $\tau_{VI} \ll \tau_c$) is controlled by effective ionization coefficient $\alpha_{eff}(n_e) = (\alpha - \eta) + (k_d - \beta_{ei}) N_n(n_e) / u_e$, where $N_n(n_e)$ is the concentration of negative ions, and u_e is the drift velocity of electrons. In this case, the excess of the rate of impact ionization by the rate of sticking of electrons is compensated by the separation of electrons from negative ions by electron impact. The SSVD on this branch of the VAC is unstable with respect to spatially uniform fluctuations of the plasma parameters. However, in plasma of the SSVD, non-uniform fluctuations can grow, leading to the formation of spatial structures in the form of individual current cords with an increased electron concentration. The corresponding scenarios of self-organization and the problem of the stability of the plasma structures arising in this case are discussed in detail in the literature [158, 159]. It seems likely that, as a result of an instability developing in a falling portion of the VAC, an SSVD contraction occurs with a long duration of the discharge pulse (Fig. 4.22).

4.4.4. Mechanism of propagation of conducting channels in SF_6

The considered nonlinear mechanism of electron multiplication due to an imbalance of constants k_d and β_{ei} can even lead to the development of ionization instability in SF_6-based gas mixtures at room temperature, if, as already noted, $(E/N) \gg (E/N)_{cr}$. This is the case near the tip of a single conductive channel, growing from the cathode.

At the tip of such a channel, the reduced electric field strength significantly exceeds the value of $(E/N)_{cr}$ [44]. This leads to a significant increase of T_e and, as a consequence, a marked decrease of β_{ei}. A situation arises in which, again, as in the case of a high gas temperature, $(k_d - \beta_{ei}) > 0$. As a result, the above-discussed mechanism of the 'explosive' increase in the electron concentration is initiated and a new, plasma-filled, section of the channel is formed, ensuring its advance into the discharge gap. In this case, there is no need to involve the mechanism of stepwise ionization of the SF_6 molecules proposed in Ref. [103] to explain the propagation of conducting channels in SF_6 and its mixtures.

4.4.5. Analysis of the results

Section 4.4 investigated a fundamentally new mechanism for the development of detachable instability developing in active media of HF(DF) lasers due to electron detachment from negative SF_6^- and SF_5^- ions. Based on the analysis of the main mechanisms of formation and destruction of negative ions in SF_6 and its mixtures, it is shown that the instability arises as a result of an imbalance of the rate constants of electron separation from negative ions by electron impact and dissociative electron–ion recombination. Analytical expressions for the electron density change with time are obtained. It is shown that with an increase in discharge current, the process of developing instability is 'explosive' in nature, and an estimate of the characteristic time of 'explosion' is given. By laser heating of gas mixtures based on SF_6 by radiation from a pulsed CO_2 laser, it was initiated, with the aim of experimental study, to develop the detachment instability in large volumes. The connection of this process with the phenomenon of spatial self-organization (formation of current cords) in the plasma of the SSVD in pre-irradiated SF_6 and its mixtures is shown. The mechanism of propagation of a single incomplete channel due to the process of electron detachment from negative ions by electron impact is considered.

Conclusions to Chapter 4

1. The phenomenon of self-organization of SSVD during laser heating, which leads to the formation of quasi-periodic current structures in the volume of the discharge gap, was discovered. It is established that heating (a strong population of the vibrational states of SF_6) leads to discharge stratification (the appearance of filamentary structures in the heating zone) and the development of plasma instability (SSVD contraction (SIVD)). Direct experiments have shown that heating a gas can lead to the formation of a plasma instability in the volume of the discharge gap, and not from the electrodes.

2. By comparing the experimental and calculated oscillograms of the voltage and current of the SIVD, it was shown that the discharge contraction in the working mixtures of non-chain HF (DF) laser can be caused by a decrease in the electron-ion recombination process rate due to temperature increase, which ceases to compensate for the increase in electron concentration due to processes of electron impact electron impact from negative ions.

3. For the first time, an independent volume discharge in SF_6 and mixtures based on it under conditions of shock-wave perturbations caused by gas heating in the discharge gap by a pulsed CO_2 laser was investigated. A number of previously unobservable effects were discovered, in particular, the barrier effect and the discharge 'slip' effect on the boundary of the heated and cold gas zones, outwardly similar to the development of a sliding discharge on the surface of a dielectric placed between metal electrodes.

4. It has been established that, at the boundaries of shock-wave perturbations in the SSVD plasma, regions with uncompensated surface charge form in the gas, which substantially distorts the distribution of the electric field in the gap and changes the structure of the discharge. Using the technique of a delayed nanosecond discharge, the dynamics of the formation of shock-wave perturbations

of a gas caused by pulsed heating of a gas using a high-power pulsed CO_2 laser was studied in detail.

5. It has been established that $(E/N)_{cr}$ in undissociated SF_6 monotonously increases with T_g due to an increase in the rate of attachment of electrons to vibrationally excited SF_6. The values of $(E/N)_{cr}$, obtained when determining the temperature by two different methods, are in good agreement with each other.

6. Based on the analysis of the literature and experimental studies of the temperature dependence of the critical electric field in SF_6, it is shown that the dependence of $(E/N)_{cr}$ in SF_6 on the plane of the parameters T_g and t (T_g is the temperature, t is the time from the moment of pulse heating of the gas) has a singularity which topologically seems to be a feature of only one type – 'assembly crash'. An interpretation of this phenomenon is given.

,

5

Powerful pulse and pulse-periodic non-chain HF(DF) lasers

In this chapter, the dissertation presents the results of studies of the characteristics of non-chain HF(DF) lasers initiated by SSVD. The problems of creating non-chain HF(DF) lasers of a kilojoule level and P–P lasers operating with a high pulse repetition rate are discussed. The main results presented in this chapter were obtained in [20, 160–165].

5.1. Non-chain electric-discharge HF(DF) lasers with a short discharge pulse duration and a small volume of active medium

The results presented in Chapters 2 and 3 clearly indicate that a uniform volume discharge in non-chain HF(DF) laser working mixtures can be realized without the use of special pre-ionization devices. This circumstance greatly simplifies the design of the laser and opens up great prospects for increasing the working volume of the laser in order to further increase the energy characteristics of the laser.

However, in installations with volumes of the active medium less than 1 litre with a relatively small cathode area in the absence of forced discharge initiation, there is a significant variation in the voltage amplitude and delay time of the pulsed breakdown of the discharge gap (DG), which is especially undesirable in the P–P mode of operation [71]. Therefore, in this case, it is advisable to forcibly initiate a gap breakdown, for example, with a low-current spark located either to the side of the DG or in the hole on the cathode [20]. In principle, this mode is similar to the 'photo triggered discharge' mode [166, 167]. But in standard 'photo triggered discharge' schemes used in excimer lasers [167], powerful gap illumination is necessary, since radiation performs two functions: 1) initiates a breakdown, 2) creates the necessary initial concentration of electrons in a gaseous medium. In SF_6 mixtures with hydrocarbons a powerful illumination is not needed, since a distinctive feature of the SSVD is that after a local breakdown of the gap in an arbitrary place, the discharge, as shown above, spreads over the entire surface of the cathode. This means that the local illumination of the cathode with a low-current spark should be sufficient to stabilize the electrical and output characteristics of the HF(DF) laser. In this section, we will consider the features of the operation of a HF(DF) laser with an aperture of up to 5 cm.

5.1.1. Description of the experimental setup

The laser used flat Al electrodes with dimensions of 20 × 80 cm (anode) and 7 × 60 cm (cathode), rounded around the perimeter with a radius of 1 cm and spaced by a distance of $d = 5$ cm. The surface of the cathode was sandblasted. To obtain the SSVD in the form of SIVD, the Fitch generator circuit (Fig. 5.1.) which is standard for small aperture lasers, with a capacitor capacitance of C_1, $C_2 = 0.1$ µF and a maximum charging voltage of 50 kV was used. From the side, the DG was illuminated by a spark bounded by two resistances $r = 5$ kΩ, connected directly to the electrodes. The spark was located at a distance of ~5 cm from the edge of the cathode symmetrically with respect to the electrodes. The operation of the laser in the P-P mode with a frequency of up to 10 Hz was provided by placing ten fans in the chamber. The laser worked on SF_6 mixtures with C_2H_6 and C_6D_{12} at a total pressure of 45–70 mmHg. In most experiments, a resonator was used, formed by an Al mirror with a radius of curvature of 20 m and a plane parallel BaF_2 plate.

Note that, in contrast to the typical photo triggered discharge 'system, in our case, the initiation of a breakdown occurs automatically as soon as the voltage across the gap exceeds a certain critical value. By separating the illumination circuit from the laser pumping circuit, it is possible to initiate a breakdown at an arbitrary time. The pulse shape of the HF laser was recorded by a pyroelectric receiver with a time resolution of ~20 ns.

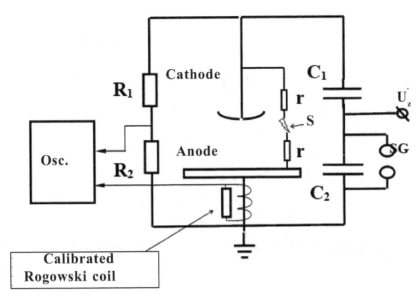

Fig. 5.1. Electric circuit of the installation for studying the characteristics of small-aperture non-chain HF(DF) lasers initiated by SIVD: U_z = 36÷50 kV; C_1 = C_2 = 100 nF; SG – arrester under pressure; R_1, R_2 – high voltage divider. S – the spark to initiate the breakdown of the discharge gap, r = 5 kΩ.

5.1.2. Experimental results

Figure 5.2 shows typical oscillograms: an HF laser pulse, discharge current and voltage (curves 1, 2 and 3, respectively). As can be seen from this figure, thanks to photoinitiator, the breakdown of the DG occurs at the leading edge of the voltage, the maximum of the generation pulse is somewhat delayed relative to the current maximum. When the initiating spark was turned off, the spread of the amplitude of the breakdown voltage of the DG from pulse to pulse reached 20%, which, respectively, caused a 15% spread of the laser radiation energy values. Note that with such a long current pulse duration (τ_{dis} > 400 ns), it was not possible to obtain a stable SSVD (SIVD) in H_2:SF_6 mixtures – the discharge had a contracted form. Only in C_2H_6:SF_6 mixtures in the experimental conditions described above was a volume discharge realized. Figure 5.3 shows the dependence of the output energy of the laser W_{out} (generation on HF) on the energy W introduced into the SIVD plasma for mixtures with different contents of C_2H_6. From Fig. 5.3 it can be seen that in mixtures with a ratio of C_2H_6:SF_6 = 1.5:22 and 2:22, the output energy grows almost linearly with increasing energy input. Under

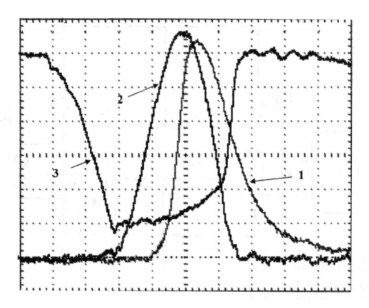

Fig. 5.2. Oscillograms of laser generation pulses (1), discharge current (2) and voltage (3) on DG; current 3 kA / div, voltage 10 kV / div, sweep 100 ns / div. A mixture of 66 mmHg SF_6 and 6 mmHg C_2H_6.

the experimental conditions, a mixture with a C_2H_6:SF_6 = 1.5:22 ratio turned out to be optimal, at which the maximum value of the generation energy W_{out} = 8 J at electrical efficiency = 3.2% was obtained. The discharge volume estimated from the laser radiation imprint on a heat-sensitive film was ~1.5 l, which corresponds to the specific energy contribution to the plasma ~220 J/l. The decrease in W_{out} with an increase in W for mixtures with a low content of C_2H_6 (a mixture of C_2H_6:SF_6 = 1:22) is associated with a loss of stability of SSVD at high energy input. Indeed, when working on this mixture with energy inputs of ~200 J/l, bright incomplete channels were observed in the discharge chamber, growing from the edge of the cathode, which sometimes blocked the DG. For mixtures with a higher C_2H_6 content, a decrease in the efficiency of the laser with increasing W did not occur as long as the discharge remained stable and the length of the incomplete plasma channels did not exceed $d/2$. This gives grounds to expect an increase in efficiency with an increase in the interelectrode distance, due to an improvement in the uniformity of the SIVD in the form of SSVD as a result of a more complete overlap of diffuse channels.

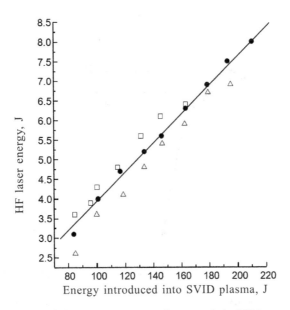

Fig. 5.3. Dependence of the output energy of a non-chain HF laser on the energy deposited in an SIVD plasma for mixtures with different ethane contents: $\square$ – $C_2H_6:SF_6 = 1:22$; $\bullet$ – $C_2H_6:SF_6 = 1.5: 22$; $\triangle$ – $C_2H_6:SF_6 = 2:22$ (SF_6 pressure was 66 mmHg).

In the system of electrodes under study, there is a large amplification of the electric field at the edge of the gap. In gases such as CO, CO_2, air, N_2, this leads to the discharge of the discharge to the edges of the gap [71]. In mixtures of SF_6 with hydrocarbons (carbondeuterides), due to the specifics of SIVD, this does not occur; even starting at the edge, SIVD is forced into the depth of the DG due to the self-organizing processes of current density distribution in the diffuse channels. However, as shown in section 2.5 of the dissertation, when studying the dynamics of propagation of a SIVD, even in an electrode system with a sharply inhomogeneous field (plane–blade geometry), separate diffuse channels form on the cathode–blade and it takes some time for the next ones to appear, and if ones force the input energy, instead of the formation of new channels, it is possible to observe the effect of current oscillations in the diffuse channel, which in turn may provoke the development of instability. Under conditions when the duration of the current pulse exceeds 150 ns and the specific energy input in SF_6-based gas mixtures with hydrocarbons exceeds 80 J/l, SIVD has time to fill the entire DG by the end of the discharge pulse (it is assumed that there are microinhomogeneities in the cathode).

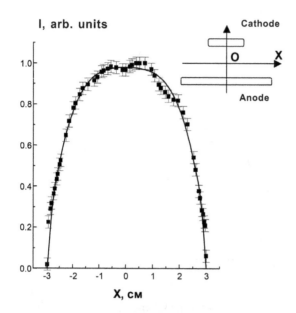

Fig. 5.4. The intensity distribution of the emission of a SIVD plasma in a plane parallel to the surfaces of the electrodes and passing through the optical axis.

Figure 5.4 shows the distribution of the intensity of the emission of a SIVD plasma in a plane parallel to the surfaces of the electrodes and passing through the optical axis. From Fig. 5.4 it can be seen that the maximum intensity of the emission of the SIVD is located on the optical axis. In a similar way, the radiation energy is distributed over the laser aperture, i.e. the edge amplification of the electric field does not have a significant effect on the distribution of the output laser radiation. It should be noted that the observed distribution of the energy input is observed when the cathode surface has undergone special treatment in order to create asperities on it. If the cathode surface was polished, then without uniform illumination of the cathode, a non-uniform SIVD was obtained, the energy release was localized in places with the greatest amplification of the electric field in the DG.

Thus, after treating the cathode surface in order to create asperity of about 50 μm in it in SF_6 mixtures with hydrocarbons, local illumination of the cathode is sufficient to obtain a discharge uniformly distributed throughout the volume, and the presence of areas with a high edge amplification of the electric field practically does not impair the stability of the SIVD it has little effect on the energy distribution of laser radiation over the aperture. Therefore, you can use flat electrodes with a small radius of curvature at the edge.

5.2. Measurement of the divergence of the radiation of a non-chain HF (DF) laser initiated by an SIVD

For most applications of high-power HF(DF) lasers, it is especially important to control the quality of laser radiation (its angular directivity). The registration of the space–time distribution of the intensity of radiation from high-power pulsed lasers is essential to study the transmission of laser radiation over long distances and to achieve effective interaction with targets. The laser radiation pattern generated by an optical resonator can be regulated by adaptive and non-linear optics methods [168]. To form the required spatial distribution of the intensity of radiation from wide-aperture laser systems using adaptive optics, it is necessary to measure the shape of the surfaces of the wave function using WF (wave front) sensors. Thus, for practical applications of non-chain HF(DF) lasers with given power parameters, there are two more acute problems: obtaining laser radiation with a good radiation pattern (i.e. with the minimum possible divergence), and the need to control the WF laser, in order to adapt the laser radiation pattern using adaptive optics methods. We consider these tasks sequentially.

The divergence of the radiation of a non-chain HF(DF) laser, initiated by a SIVD, is limited by the appearance of optical inhomogeneities of the active medium, which arise during the discharge and generation of radiating HF(DF) molecules. The main reasons for the appearance of optical inhomogeneities may be compression and rarefaction waves arising in the near-electrode zones, medium disturbances due to the effect of radiation self-action in the active medium, as well as inhomogeneity of the energy input. In non-chain HF(DF) lasers, the pumping times are rather short (~400 ns), therefore, the gas-dynamic density perturbations can be neglected, since they do not have time to develop during these times. Perhaps, these disturbances will play a significant role in the operation of the laser in the P–P mode with a high pulse repetition rate. For pulsed non-chain HF(DF) lasers, it seems that the main factor determining the divergence of radiation is medium inhomogeneities, due to the fundamentally jet structure of the SSVD in SF_6-based mixtures. This section presents the results of studying the divergence of laser radiation, described in Section 5.1.1, when it is operated on an $SF_6:C_6D_{12}$ mixture (lasing on a DF molecule).

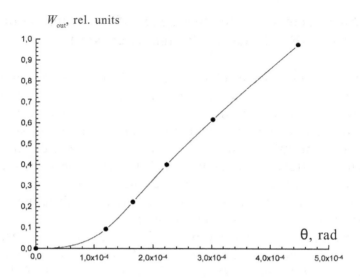

Fig. 5.5. The distribution of the radiation energy of a DF laser W_{out} over the angle θ in the far zone.

5.2.1. Methods of measuring the radiation divergence and the experimental results

The divergence of the laser radiation was measured using an unstable telescopic resonator with an increase factor of $M = 3$. The experiments were performed on a laser with an interelectrode distance of $d = 5$ cm, described in section 5.1.1 of this book. To eliminate the influence of the near-electrode areas (areas with cathode plumes), the laser aperture in these experiments was limited to 4 cm in diameter. The divergence of the radiation was measured using the focal spot method using a mirror wedge as described in detail in [169].

The results of measurements of the divergence of the DF laser radiation are shown in Fig. 5.5, which shows the distribution of radiation energy by angle θ in the far zone. As can be seen from this figure, the radiation divergence at a level of 0.5 was $θ_{0.5} = 2.8 \cdot 10^{-4}$ rad that for a 4 cm aperture there corresponds to two diffraction limits. Some improvement in this parameter can be expected with an increase in the laser aperture, since the discharge uniformity should be improved with an increase in the interelectrode distance due to a more complete overlap of the diffuse channels.

5.3. Control of HF (DF) lasers using Talbot interferometry

Of the many methods of optical control of waveforms [170,171], wavelength sensors based on the shadow Hartman method are advertised to measure the shape of surfaces of waveforms in the range of the spectrum of 3–5 μm. However, the approach [172], which is based on the Talbot effect – the Hartman interference-analog method, is more attractive for analyzing the WF of wide-aperture high-power lasers. In the Talbot effect, flat coherent waves of laser radiation sources decompose on a two-dimensional array of arbitrary in shape but periodically arranged holes into spatial harmonics propagating at multiple angles, which as a result of interference in the near Fresnel zone reproduce the intensity distribution on the array at multiple distances

$$L_n = 2p^2 \, n/\lambda, \tag{5.1}$$

where $p \gg \lambda$; L_n is the distance from the grating to the playback plane, p is the grating period, λ is the radiation wavelength, $n = 1$, 2, 3 ...

The intensity distribution in the playback planes has a periodic structure similar to the grating, if the WF is flat or spherical. When the WF is flat, the periods in the reproduction plane of the grating coincide with the grating periods. This property of the effect is advantageous to use when aligning a laser resonator with flat mirrors, since the coincidence of the periods in the intensity distribution on the screen in the reproduction plane with the grating periods indicates the flatness of the WF. With convex WF, periods increase, and with concave, they decrease. If the WF is not flat or spherical, then periodicity is violated in the playback plane. An increase in the period in a certain region of the beam aperture indicates the appearance of an optical inhomogeneity in the region of the type of a scattering lens, and a decrease in the focusing lens.

The light spots of the interferograms (talbograms) on the receiving screen in the Talbot interferometry method, in contrast to Hartmangrams, are contrasting, because the width of the spots borders is determined by the diffraction of the WF on the aperture of the grating, which does not limit the laser beam aperture. The measurement of the shift of the coordinates of contrasting spots in the intensity distribution of talbograms distorted by WF in the playback planes is the basis of the HF sensor. The local slopes of the

WF $\Delta\gamma$ are measured in the reproduction planes. In the approximation of the optical wedge

$$\Delta\gamma = \Delta r/L, \ \Delta r = [(\Delta x)^2 + (\Delta y)^2]^{1/2}/L, \tag{5.2}$$

where Δx, Δy are the displacement coordinates of the spots in the rectangular coordinate system. Normal deviation of the WF $\Delta z = p\Delta\gamma$. The radius of curvature of the WF R is determined in a parabolic approximation by the change of periods Δp in the reproduction plane

$$R = pL /\Delta p, \tag{5.3}$$

Δp is the period change in the playback plane.

The special feature of Talbot interferometry is that in order to obtain information on the shape of the wave function, a detailed measurement of the spatial distribution of the radiation intensity is not required. It is sufficient to measure only the coordinates of the centres of the contrasting spots of radiation on talbograms, and the size of the spots can be chosen small compared with the period. This reduces the requirement for the dynamic range of the sensitivity of the screens and simplifies the automatic reading of coordinates. The transfer of coordinates to the PC memory for plotting the waveform can be performed by electronically scanning talbograms or using photosensitive matrices. Currently, for the optical range, there are reliable and cheap photosensitive CCDs with more than 106 elements.

The main purpose of the research presented in this section of the book was to search for wide-aperture receiving screens for recording the spatial distribution of the laser radiation intensity and the shape of waveforms of pulsed electric-discharge HF(DF) lasers using the interferometry method based on the Talbot effect, as well as developing a non-chain HF (DF) laser initiated by SIVD, with high quality laser beam [162].

5.3.1. Experimental setup and experimental technique

5.3.1.1. Selection of screens for visualization of non-chain HF(DF) laser radiation

The main criterion for choosing the screen material was the possibility of obtaining the processed interferogram in a direct laser beam (unfocused) after its passage through the periodic grating. In accordance with this requirement, the screens must have a sufficiently high sensitivity to the effects of laser radiation. We investigated the following types of screens.

1) Screens based on heat-sensitive paper (in this case, paper for fax machines and copy paper was used). The image obtained on thermosensitive paper due to a change in its colour at the sites of radiation exposure was then scanned and entered for further processing in a PC.

2) Phosphorus screens operating on the effect of temperature quenching of luminescence (when the phosphorus is irradiated with UV radiation) by IR laser radiation. A phosphorus based on ZnS–CdS polycrystals doped with Ag and Ni was used. The resulting image was photographed with a digital camera and entered into a PC.

3) Metallized (Al coating) polyethylene film with transparency in visible light of ~50%. In this case, the interferogram was obtained as a result of the destruction of the coating at the maxima of the laser radiation intensity. The resulting image was also scanned and entered into a PC for processing.

4) Graphite screens. The screens were made by applying on paper or a thin film of graphite powder. The glow of graphite under the influence of laser radiation was photographed with a digital camera, the image was taken in a PC.

In many cases, a digital camera was used to register interferograms, which required the development of a special scheme for synchronizing camera shutter opening times and laser starting. This scheme is shown in Fig. 5.6. The synchronization scheme of switching on a digital camera with the launch of a laser: FA – a digital camera; PD – photodiode; GPG – delay pulse generator; SS – system of start of the generator of pulse voltage; PVG – pulse voltage generator for pumping an HF(DF) laser.

Start is carried out by a flashlight of the digital camera. The flash is detected by a silicon photodiode, the signal from which is fed to the DPG. DPG with a delay of ~150 ms, a pulse is generated that goes further to the PVG launching system for pumping a laser.

Fig. 5.6. The synchronization scheme for turning on the digital camera with the laser start: DC – digital camera; PD – photodiode; DPG – delayed pulse generator; SS – start-up system for a pulse voltage generator; PVG is a pulse voltage generator for pumping an HF (DF) laser.

The need for such a large launch delay is due to the peculiarity of the digital camera's flash lamp, in which two backlight pulses are formed, separated by 150 ms in time, the shutter of the camera opens only after the second flash.

5.3.1.2. Experimental setup on registration of interferograms

The general scheme of the experiment to study the possibility of imaging radiation of non-chain HF(DF) laser with various screens (more precisely, visualization of interferograms obtained by the Talbot-interferometry) is presented in Fig. 5.7. The laser beam passes through a periodic grating (0.1 mm thick, the grating period is $p = 1$ mm) and falls on the surface of the screen installed at a distance $L = 2 \cdot np^2/\lambda$. The synchronization scheme for turning on the digital camera with the laser start: DC – digital camera; PD – photodiode; DPG-delayed pulse generator; SS – start-up system for a pulse voltage generator; PVG is a pulse voltage generator for pumping an HF (DF) laser $2 \cdot np^2/\lambda$ (flat WF), where $n = 1, 2 ..., p$ is the the lattice period, λ is the wavelength (see formula (1)). Most of the experiments were carried out with $n = 1$. The value of L was calculated for the most intense line in the spectrum of a HF laser with a wavelength of $\lambda = 2.7$ μm. Accordingly, the calculated value was $L = 74$ cm for $n = 1$. One of the objectives of the experiments was to ascertain the possibility of obtaining a contrast interferogram (talbogram) under conditions where a large number of lines are present in the laser emission spectrum. In this case, the interference peaks for the selected line are superimposed on the 'gray' background created by other laser generation lines.

When using the phosphorus and graphite screens, the image on the screen was photographed with a digital camera, which was installed either at an angle to the optical axis to remove the screen from the side of a laser beam falling on it, or on the optical axis from the back side of a translucent screen in visible light (phosphorus screen, graphite screen on vinipros plastic film).

During the experiments, the energy density of the laser radiation incident on the periodic grating varied within $W = 0.12 \div 0.3$ J/cm^2.

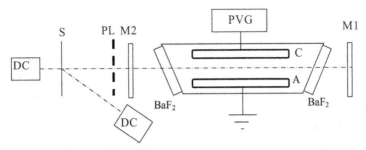

Fig. 5.7 Experimentap setup: M_1, M_2 – HF (DF) laser mirrors; BaF_2 - BaF_2 windows; PG – periodic grating; S – screen; DC - digital camera; PVG – pulse voltage generator; C is the laser cathode; A is the laser anode.

5.3.1.3. Electric-discharge HF(DF) laser for testing the technique of controlling WF by Talbot interferometry

The main problem in creating an electric-discharge HF(DF) laser is to obtain a stable SSVD at sufficiently high (up to 200 J/l) specific contributions of electrical energy to the gas-discharge plasma. The possibility of implementing a stable discharge, first of all, is determined by the performance of the electrode system and the high-voltage PVG, which should provide the extremely short duration of the discharge current for a given energy. In the created HF(DF) laser, the discharge volume was $V = 6\times6\times80$ cm^3. Flat Al electrodes on the perimeter were rounded with a radius of $r = 2$ cm. The surface of the cathode was subjected to sandblasting in order to ensure the sustainability of the discharge. The electrode system was placed in a glass-epoxy discharge chamber with an internal diameter of 500 mm and a length of 1200 mm. At the camera ends, BaF_2 windows were installed at an angle of ~5° to the optical axis. The working mixture (a mixture of SF_6 with hydrogen or deuterium donors) was prepared directly in the discharge chamber. The inlet of the working gases was carried out in a pre-evacuated backing pump and washed SF_6 chamber. The ratio of the chamber volume to the discharge volume was ~80, which allowed producing up to 50 discharges without changing the working mixture. To sustain the initiation of breakdown of the DG at a given voltage parallel to the DG, a spark gap bounded by the resistance $R = 5$ kΩ was connected to the side of the electrodes. Such a spark cannot, in principle, carry out the bulk photoionization of gas in the chamber, but due to the photoelectric effect on the cathode, it helps to stabilize the amplitude and breakdown time of the DG when a high-voltage pulse from the PVG is applied to it, which is very important to ensure the reproducibility of the energy and power of the laser radiation pulse from shot. This scheme is shown in Fig. 5.8. It works as follows. The Arkad'ev–Marx five-stage power-supply device with a capacitance at impact C_1 = 50 nF, shown in Fig. 5.8 for simplicity, with a combination of capacitor C_1 and discharger P_0, charges capacitor C_2 = 50 nF for ~3.6 μs to a voltage of 100÷130 kV. Then the starting system ignites the nitrogen-filled discharger P, and the voltage is applied to the peaking capacitance C_3 = 30 nF and DG. Due to the low-inductance connection in parallel with the DG of an intensifying capacitor $C_3 < C_2$, in this case, the effect of shortening the discharge current is achieved.

In the process of finalizing the PVG, the capacitance value of the capacitor C_3 was chosen experimentally in order to ensure the matching of the wave resistance of the circuit with the resistance of the gas-discharge plasma. The optimal value of C_3, at which the stable SIVD is realized in a wide range of energy inputs to the gas-discharge plasma, was C_3 = 30 nF. PVG, shown in Fig. 5.8, was assembled from a combination of capacitors KMKI-100-0.1, PCF-50-0.03 and IK-100-0.25. The duration of the energy release in the gas-discharge plasma was ~250 ns. With such a duration of the discharge current, the discharge in the SF_6 mixtures with hydrocarbons or

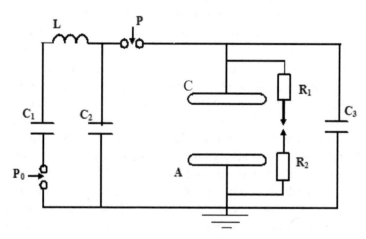

Fig. 5.8. Electrical diagram of a pulsed voltage generator (PVG) for pumping an electric-discharge HF (DF) laser: C_1, P_0 is a Arkadyev–Marx pulsed voltage generator with a capacitance at impact C_1 = 50 nF; L = 52 μH – inductance, which includes the inductance of the Arkad'ev–Marx generator; C_2 = 50 nF; C_3 = 30 nF; P – nitrogen-filled discharger; R_1 = R_2 – resistance to limit the spark current; A – the anode; C – cathode.

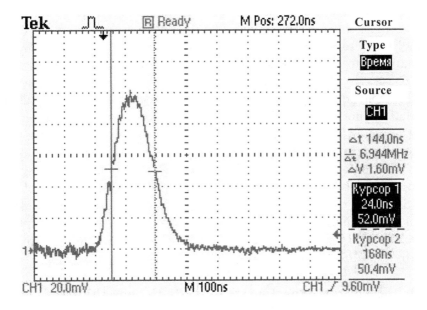

Fig. 5.9. Typical oscillogram of HF laser radiation pulse. The time scale is 100 ns/div.

carbon deuterides was stable at specific contributions of electrical energy up to 200 J/l. With specific contributions of less than 90 J/l, stable SSVD is also realized when H_2 is used in a mixture. For good reproducibility of the laser radiation energy from pulse to pulse, a stable triggering of the spark gap P in the circuit in Fig. 5.8 with respect to Arkad'ev–Marx PVG dischargers is necessary. This stability was achieved due to the use of a multichannel block for igniting arresters assembled on the basis of TGI-1-1000/25 thyratrons and K15-10 capacitors in the laser start-up system. To reduce the variation in the response time of the arresters, the starting impulse from the launching system was simultaneously applied to the ignition of the first three stages of the Arkad'ev–Marx generator.

The laser resonator was formed by a flat Al mirror and an output mirror with a dielectric coating on a CaF_2 substrate. The reflection coefficient of the output mirror at a wavelength of $\lambda = 3$ μm was 30%. In the absence of the intracavity diaphragm on the laser beam prints on thermal paper, parallel bands were observed associated with the reflection of radiation from the surfaces of the electrodes. When the output mirror was replaced with a BaF_2 plate, the laser energy did not decrease, but the relative brightness of the bands increased, which indicates the removal of a significant fraction of the energy in the amplification mode with the reflections from the electrodes. To suppress this negative effect, two round diaphragms (at a distance of ~10 cm from each of the mirrors) with a diameter of 5 cm were installed into the cavity of a modified laser layout.

The laser radiation energy was measured using an E-60 radiation energy converter, which was installed in a direct beam. The maximum radiation energy of the HF laser, obtained before the installation of restrictive orifice plates inside the resonator and when installing the resonator mirrors directly at the ends of the gas-discharge chamber, was 12 J, with a technical efficiency of 3.2%. The radiation energy of a modified laser model (after intracavity diaphragm) was $E = 4.5$ J on an $SF_6:C_2H_6 = 10:1$ mixture used in most visualization experiments.

The laser pulse shape was recorded by a photovoltaic laser power receiver with a time resolution of ~1 ns. Figure 5.9 shows a typical oscillogram of a HF(DF) laser radiation pulse.

5.3.2. Results of experiments to study the possibility of visualizing the radiation of non-chain HF(DF) laser by various screens

Numerous experiments with screens based on heat-sensitive paper did not give a positive result. The sensitivity of the screens to the effects of laser radiation in a direct beam (unfocused) turned out to be insufficient to register the interference pattern. Contrast interferograms (talbograms) were obtained using a metallized film as a screen. For reliable recording of talbograms, the energy density of the laser radiation incident on the periodic grating should have been greater than $W = 0.2$ J/cm². A scanned image of a laser beam print on a screen of metallized film, mounted at a distance of $L = 73$ cm from the periodic grating, is shown in Fig. 5.10. As can be seen from this figure, the image is contrasting, despite the large number of lines in the laser generation spectrum, which makes it possible to determine the coordinates of the centers of the spots on the interferogram (talbogram) and construct two-dimensional picture of the WF laser beam. The disadvantage of the screen based on the metallized film is a rather high threshold of optical destruction of the metal layer, as well as the 'disposability' of such a screen.

Graphite screens turned out to be much more convenient, starting to glow already at a fairly low density of energy incident on the periodic grating, $W \approx 0.12$ J/cm². Figure 5.11 shows interferograms obtained on a paper screen coated with graphite powder at $L = 73$

Fig. 5.10. Scanned image of a laser beam print on a metallized film obtained at a

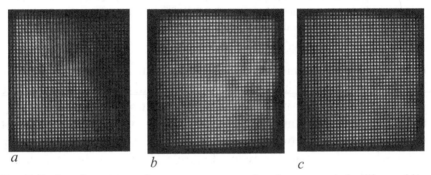

a   *b*   *c*

Fig. 5.11. Interferograms taken on a screen made of paper coated with graphite powder: (a), (b) - the laser resonator is misaligned; (c) - mirrors are adjusted.

cm. They reflect the alignment process of the resonator: Fig.11a, b – the resonator is misaligned, Fig.11c – the resonator mirrors are well exposed. From Fig. 5.11 it can be seen that misalignment of the resonator leads to 'blurring' of the interference pattern, and if the mirrors are properly installed, the interferogram becomes bright and contrast. Figure 5.12 shows interferograms taken on the same screen at L = 74 cm, which illustrate the effect of atmospheric disturbances on the WF of laser radiation: Figure 5.12a – the atmosphere in the path of the laser beam is not disturbed; Figure 5.12b – atmospheric disturbance by heat flux due to the fact that a candle burns behind the output mirror at a distance of ~1 m from the periodic grating and 20 cm below the optical axis of the system.

From Fig. 5.12 it can be seen that atmospheric disturbances due to heat flow significantly distort the WF of a laser beam. Screens based on a film of vinipros (plastic film) coated with graphite powder

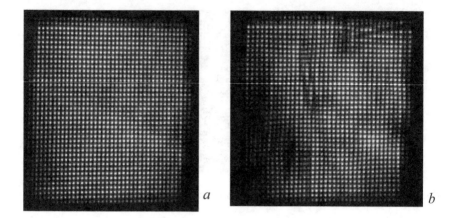

Fig. 5.12. Interferograms obtained on a screen made of paper coated with graphite powder: (a) – the atmosphere in the path of the laser beam is not disturbed; (b) – the atmosphere is perturbed by the heat flow.

turned out to be the most convenient for visualizing interferograms. These screens are transparent to visible light, so the luminescence of graphite under the influence of laser radiation can be detected by a digital camera mounted behind the screen on the optical axis of the system, which avoids picture distortions caused by the shooting angle when using opaque screens. Figure 5.13 presents interferograms obtained on a translucent graphite screen, which illustrate the change in the interference pattern with a change in the distance L between the screen and the periodic grating. From Fig. 5.13 it can be seen that the deviation of L from the optimal value $L = 73 \div 74$ cm leads to noticeable distortions of the interference pattern.

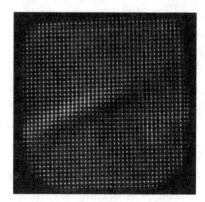

a

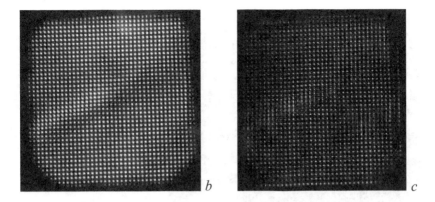

Fig. 5.13. Interferograms with different values of L, taken on a semi-transparent graphite screen: (a) $L = 59$ cm; (b) $L = 67$ cm; (c) $L = 85$ cm.

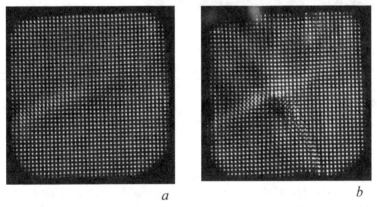

Fig. 5.14. Interferograms obtained on a semi-transparent graphite screen: (a) the atmosphere in the path of the laser beam is not disturbed; (b) atmospheric disturbance by heat flow.

Figure 5.14 shows interferograms illustrating, as in Fig.5.12, the influence of atmospheric disturbances on the WF by the heat flux on the beam path. Figure 5.15 shows for comparison the interferogram taken at a distance between the semitransparent screen and the periodic grating $L = 146$ cm ($n = 2$).

From Fig. 5.15 it is seen that an increase in the sequence number n leads to a decrease in the brightness of the interference pattern. It is significant that at the same time the extreme rows of luminous points disappear on the interferogram, i.e. the total amount of information on the wave function is reduced, which can be obtained from the interferogram. Therefore, it is advisable to remove the interferogram when $n = 1$.

Fig. 5.15. Interferogram obtained on a translucent graphite screen with $L = 146$ cm ($n = 2$).

Thus, as a result of experimental studies, the possibility of visualization of HF (DF) radiation of a laser passing through a periodic grating with various screens was shown. Despite the presence of a large number of lines in the HF(DF) laser emission spectrum, contrasting interferograms were obtained, which indicates that these screens can be used to control non-chain HF(DF) laser WF using Talbot interferometry.

In conclusion, it should be noted that the quality of the laser beam created by the HF (DF) laser layout is quite high, as evidenced by the very possibility of obtaining contrast and undistorted talbograms at distances corresponding to the 3rd reproduction plane ($L3$). Without special optimization of the laser resonator, the talbograms were already distorted in the first plane of playback, and the image in the second plane was so blurred that it was impossible to determine the coordinates of individual points. In particular, when the resonator mirrors were attached directly to the ends of the discharge chamber, and they also sealed the chamber, a strong distortion of the laser beam WF due to elastic stresses in the mirrors was observed.

5.3.3. Analysis results

A model of a non-chain electric-discharge HF(DF) laser with high radiation quality was created and tested. A stable SIVD was obtained in SF_6:H_2 mixtures with specific input deposits up to 80 J/l and in

$SF_6{:}C_2H_6$ mixtures up to 200 J/l. The radiation energy of an HF (DF) laser was E = 4.5 J in a beam 5 cm in diameter. In order to quickly capture the interferograms, a scheme was developed for synchronizing and starting the laser from a flash lamp of a digital camera.

The possibility of using various screens for visualization of HF (DF) laser radiation has been investigated. Contrast interferograms were obtained on graphite screens, thermal film and screens based on metallized polyethylene film after passing a laser beam through a periodic grating. The high quality of the interferograms makes it possible to use the studied screens in the Talbot interferometry method for monitoring the HF radiation of a non-chain HF(DF) laser. According to the combination of characteristics, graphite screens with high sensitivity to laser radiation are the most convenient for applications. The combination of these screens with modern digital photographic equipment allows you to quickly enter the interferogram to be removed in a PC for further processing. The advantage of graphite screens is also the practical absence of aperture restrictions, which makes it possible to use them to monitor high-frequency wave power powerful wide-aperture P-P-HF(DF) lasers and other IR lasers.

5.4. 400 J pulsed non-chain HF laser. Scaling capabilities of the characteristics of non-chain HF(DF) lasers excited by SIVD

With an increase in the cathode area and the volume of the active medium, there is no need for a forced initiation of gap breakdown (see Section 5.2 of this dissertation). The breakdown delay in this case becomes so small that it is not fixed on the oscillograms, i.e. breakdown occurs at the voltage front. Therefore, in installations with a large volume of active medium, in order to initiate gap breakdown, there is no need for additional devices; a fairly homogeneous SIVD forms spontaneously. Consider the problem of scaling non-chain HF(DF) laser.

5.4.1. Experimental setup

In experiments on scaling non-chain HF(DF) laser, cathodes with a size $h{\times}\ell$ = 10×50 cm were used; 15×75 and 20×100 cm, rounded around the perimeter with a radius of r = 1 cm. The surfaces of the cathodes were sandblasted. A plate with dimensions of 40–120

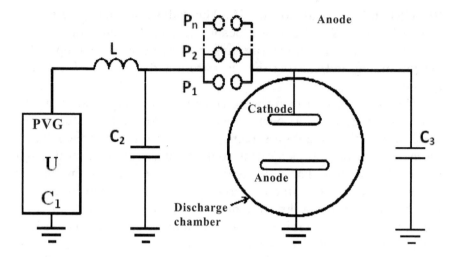

Fig. 5.16. Diagram of a laser setup for studying the possibility of scaling the characteristics of a non-chain HF(DF) laser initiated by an SIVD.

cm served as an anode. The interelectrode distance varied within $d = 10$–30 cm. The electrodes were placed in a discharge chamber, which was a glass-epoxy tube 50 cm in diameter and 140 cm long. The laser resonator was formed by a plane-parallel plate made of BaF_2 and concave Al mirror with a radius of curvature of 20 m, mounted directly on the ends of the discharge chamber. Figure 5.16 shows a typical electrical circuit of a pulsed high-voltage source used in scaling experiments to obtain a SIVD in installations with different volumes of the active medium. The PVG (pulsed voltage generator), made according to the Arkad'ev–Marx scheme, charges the capacitor C_2 through inductance L, and through parallel-fired pulse dischargers $SG_1 \div SG_n$, the voltage is applied to the peaking capacitor C_3 and the discharge gap. The ratios between the capacitors in the circuit are as follows: $C_1 = 0.8 \div 0.9\ C_2$; $C_3 = 0.35 \div 0.45\ C_2$.

5.4.2. Ability to scale the characteristics of non-chain HF lasers

If one sets some reasonable maximum voltage value of a high-voltage generator ($U_{max} \leq 500$ kV), an increase in the laser output energy and energy stored in the power source is possible only due to an increase in the capacitance $C = C_1$ of the system capacitors (proportional increase in all capacities in the circuit shown in Fig. 5.17.) with a corresponding change in the volume of the active medium V_a, in order to maintain the stability of SIVD. The simplest method of scaling the output characteristics of a laser is to increase V_a due to the active medium length l. In this case, all system parameters – the output energy W_{out}, V_a, and C vary in proportion to l, and the time parameter of the circuit $T = \pi(LC)^{1/2}$ remains unchanged, therefore, the stability of SIVD is preserved. However, for a number of well-known reasons, primarily due to an increase in the radiation load on optics, this method of scaling is not of interest.

The generation energy of a non-chain HF(DF) laser weakly depends on the mixture pressure at a given value of the specific input energy W_{in} (measured in $J \cdot l^{-1}$), unlike, for example, from a CO_2 laser [53]. Therefore, it is tempting to scale the HF(DF) laser by increasing its aperture (increasing the interelectrode distance d and the cathode transverse size h) for a given length of the active medium. In this case, the installation parameters change with an increase in the generator capacity as follows: the pump energy is $W \sim W_{st} \sim C$, the mixture pressure is $P \sim 1/d$, since the inductance value L changes insignificantly, the duration of the discharge will increase in proportion to $\sim C^{1/2}$, which will lead to a violation of the stability of SIVD. Therefore, as C increases, we must reduce the specific energy input W_{in} into the discharge plasma due to a corresponding

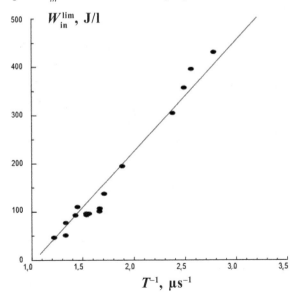

Fig. 5.17. The dependence of the value of the limiting specific energy input to the plasma W_{in}^{lim} on $1/T$.

change in V_a. Figure 5.17 shows the dependence of the value of the limiting specific energy input to the SIVD plasma W_{in}^{lim} on $1/T$, as can be seen from this figure, we can assume that $W_{in}^{lim} \times T \approx$ const is approximately. Therefore, in order to maintain the stability of SIVD, the value of V_a should increase not proportionally to C, but as $V_a \sim C^{3/2}$, regardless of whether this is achieved – by increasing the cathode transverse size or by increasing d. The observed fairly obvious patterns were indeed traced in these experiments.

It should be noted that since the pressure of the mixture P at a constant voltage changes as $P \sim 1/d$, then as V_a increases, only due to an increase in d, the specific energy input per unit volume and per molecule will increase, as C. This will increase gain of the active medium of the non-chain HF(DF) laser and may be a limiting factor in this scaling method due to the development of parasitic laser generation in the direction transverse to the optical axis. In the present experiments, this effect was not observed with d increasing to 27 cm.

The maximum generation energy of the non-chain HF(DF) laser obtained in our experiments was 407 J per HF and 325 J per DF with an electrical efficiency of 4.3% and 3.4%, respectively. The volume of the active medium was ~60 litres with an aperture of 27 cm. Oscillograms of the HF laser generation pulse, discharge current and voltage on the discharge gap are shown in Fig. 5.18. Estimation of the peak power of the laser radiation, taking into

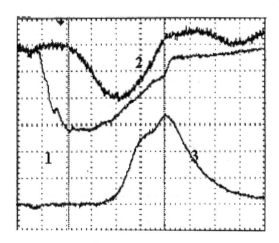

Fig. 5.18. Oscillograms of voltage pulses on the discharge gap (1), discharge current (2), and lasing pulse (3). Scan 100 ns/div.

Fig. 5.19. A photo of the appearance of a non-chain HF laser with an emission energy of 400 J.

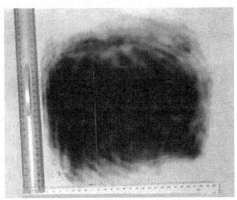

Fig. 5.20. Photograph of a laser beam print on thermal paper, obtained at a distance of 1 m from the output mirror.

account oscillograms of lasing in Fig. 5.18, gives 1.6 GW per HF and 1.2 GW per DF. Figure 5.19 shows a photograph of the appearance of the installation.

Figure 5.20 shows a photograph of a laser beam print on thermal paper obtained at a distance of ~1 m from the output mirror. For scale, here are applied rulers with a length of 30 cm. As can be seen from Fig. 5.20, the radiation density is maximum in the center, despite the fact that SIVD was ignited in the discharge gap with a high edge amplification of the electric field. This fact also confirms the results of studies of the distribution of HF laser radiation power over the aperture.

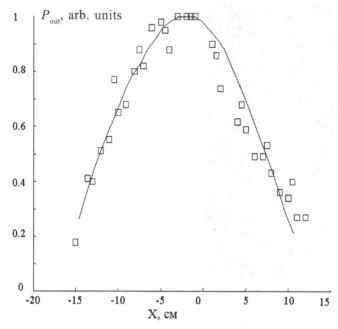

Fig. 5.21. Power distribution of HF laser radiation (P_{out}) in a plane parallel to the surfaces of the electrodes and passing through the optical axis. The interelectrode distance was $d = 27$ cm, a schematic representation of the DG is shown in Fig. 5.4.

Figure 5.21 shows the distribution of HF laser radiation power (P_{out}) in a plane parallel to the surfaces of the electrodes and passing through the optical axis. From Fig. 5.21 it can be seen that the maximum radiation power of the HF laser is on the optical axis. The photo of the SIVD obtained at the installation with a 27 cm interelectrode distance is shown in Fig. 5.22. From this figure, it can be seen that the maximum of the emission intensity of the SIVD is also on the optical axis (despite the high edge amplification of the electric field in the discharge gap).

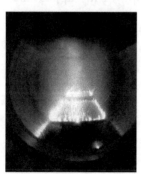

Fig. 5.22. Photograph of SIVD in a non-chain HF laser, interelectrode distance $d = 27$ cm, discharge volume ~60 l. A mixture of $SF_6:C_2H_6 = 20:1$ at a total pressure of 76 mmHg.

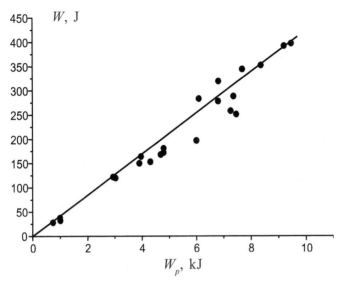

Fig. 5.23. Dependence of the radiation energy of a non-chain HF laser W on the energy W_p stored in the pump generator.

Of natural interest is the prediction of the possibility of a further increase in the laser radiation energy. Figure 5.23 shows the dependence of the output energy of the HF laser W_{out} on the energy stored in the capacitors of the high-voltage generator W_p. The graph in Fig. 5.23 shows the points obtained in installations with different volumes of the active medium. It can be seen that all the points fit well in a directly proportional relationship with an electrical efficiency of ≈4%. This allows us to predict the possibility of further increasing the output energy of the non-chain HF(DF) laser based on the methods developed by us and the creation of kilojoule or more powerful units.

In conclusion of this section, let us single out once more the conditions necessary for obtaining SIVD in large volumes of SF_6 mixtures – hydrocarbons (carbon deuterides) and creating non-chain lasers with high radiation energy:

1) small-scale, 50 μm, inhomogeneities should be applied on the cathode surface;

2) to match the wave resistance of the circuit with the resistance of the discharge plasma (for a given interelectrode distance) the pressure of the mixture should be chosen so that the discharge burning voltage determined by the conditions of breakdown of the gap in SF_6 is 2 times less than the voltage applied to the DG;

3) when increasing the electrical energy by increasing the capacitance of the generator capacitors at a given maximum voltage, the discharge volume V_a should increase, as $V_a \sim C^{3/2}$, where C is the capacity of the generator at impact. When all the listed conditions are fulfilled, it is also necessary to strive for the maximum possible reduction in the duration of the input of electric energy into the discharge plasma.

5.4.3. Pulse-periodic (P–P) HF(DF) lasers with an average power over 1 kW

The general approaches to the problem of igniting SIVD in wide-aperture HF(DF) lasers, described in the previous section, were used to create high-power non-chain HF(DF) lasers built by researchers from other institutes. To illustrate the application of the approaches presented in the previous sections of the thesis, this section briefly outlines the main results of studies of the non-chain electric-discharge HF laser with an average power of over 1.3 kW [163]. These works were carried out at the Raduga Research and Development Centre Raduga with the participation of the staff members of the Prokhorov General Physics Institute of the Russian Academy of Sciences [163, 173].

The experiments were carried out on a laser machine in which the DG of the laser was formed by two identical duralumin flat electrodes rounded around the perimeter with a radius of $r = 2$ cm. The dimensions of the flat part of the electrode surface were 15×100 cm. The electrode distance was $d = 15$ cm. The electrode serving The cathode was sandblasted to form small-scale, ~50 μm on its surface. The electrodes were placed in a glass-epoxy tube with an inner diameter of 80 cm and a length of 250 cm symmetrically about its axis. Laser working media were $SF_6:C_2H_6=20:1$, $SF_6:(C_3H_8 \div C_4H_{10})$ $=30:1$ and $SF_6:H_2=9:1$ mixtures with a total pressure of $45 \div 75$ mm Hg. The change in the working medium in the discharge volume was ensured by circulating gas in a closed loop; the flow rate in the discharge zone was 40 m/s. Special filters for the absorption of HF molecules produced in the SIVD plasma were not used, but the magnitude of the bassast volume exceeded the discharge volume by more than 2 orders of magnitude.

The laser resonator was formed by a copper mirror with a radius of curvature of $R = 20$ m and a plane-parallel KCl plate. The mirror was installed in an adjustment unit connected to the discharge chamber with a bellows. A plane-parallel plate was attached directly to the end of the discharge chamber. The pulsed voltage generator (PVG), which is used for the formation of SIVD, consisted of 4 identical sections, connected in parallel to the discharge gap electrodes by copper buses. Sections are placed in a single metal case filled with SF_6 at atmospheric pressure.

The electrical circuit of the PVD section is shown in Fig.5.24 a. The generator is assembled according to the Fitch scheme on low-inductance $C = 50$ nF capacitors with a rated voltage of 100 kV of the brand KMK-100-50. SG control arresters are filled with an SF_6:N_2=1:10 mixture with an overpressure of up to 6 atm. The device of dischargers allows to start the laser with a duration of up to 60 s at a repetition rate of discharge pulses of 20 Hz without replacing the gas in them. The charging voltage U_{ch} in the experiments varied within $U_{ch} = 50\div75$ kV. Experimentally selected inductance L_1 is used to match the circuit voltage inversion on the capacitor C with the discharge circuit. Resistive divider R_1, R_2 allows one to control

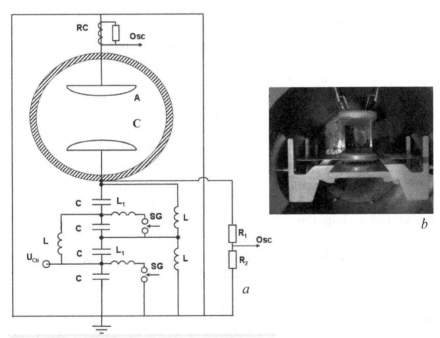

Fig. 5.24. a) Electrical diagram of a high-voltage generator (one of four sections): RC – Rogowski coil; A – anode, C – cathode; L, L_1 – inductance; SG – arrester; C – capacitor; R_1, R_2 – high-voltage resistive voltage divider; U_{ch} is the charging voltage; Osc - digital oscilloscope. b) photograph of the electrode system through the discharge chamber exit window [173].

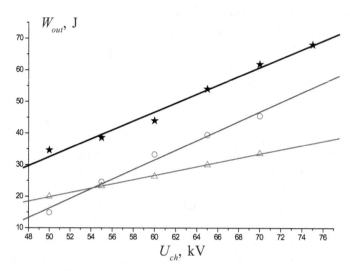

Fig. 5.25. Dependences of the laser generation energy W_{out} on U_{ch} using different hydrogen carriers: ★ – SF_6 mixture: C_2H_6 = 20: 1; O – SF_6:C_3H_8 = 30:1; Δ – SF_6: H_2 = 10:1 [173].

the pulse voltage on the DG. A part of the busbar passes through the Rogowski coil RC. Such a control method with a large busbar width (~1 m) does not allow for accurate measurements of the discharge current, but allows determining the temporary position of its maximum. The synchronization system provided simultaneous launch of 8 PVG dischargers with an accuracy of ±10 ns.

As a result of the work carried out in [163, 173], the possibility of implementing SIVD on SF_6 mixtures with hydrocarbons in the discharge gap with a high edge amplification of the electric field without additional measures to stabilize the discharge both in the pulsed and in the P–P mode is shown. Figure 5.25 shows the dependence of the output energy of the laser on the charging voltage obtained for mixtures with different hydrogen donors. The maximum lasing energy of an P–P laser at a pulse repetition rate of 20 Hz on an SF_6:C_2H_6 = 20:1 mixture reached W_g = 67 J. Thus, the average output laser power was ~1.3 kW [163, 173].

5.5. High-energy non-chain P–P HF(DF) lasers with solid-state pump generators

The possibilities of increasing the lasing characteristics of wide-aperture P–P non-chain HF(DF) lasers are limited by a number of technical problems. First, of course, this is a common problem for high-energy P–P TEA-lasers to ensure the necessary gas flow rate in discharge gaps with large dimensions. Another important problem is the difficulty of ensuring stable switching of high energies when operating in the P–P mode. The limited resource of gas-discharge switches is one of the main factors hindering the development and narrowing the scope of applications of electric-discharge lasers [52]. Therefore, it seems relevant to search for alternative switching methods, for example, on the basis of solid-state switches [164, 176], which provide a greater resource of PVG operation.

With a switched electrical power of up to several tens of joules and an output voltage of the generator up to ~100 kV, relatively low-voltage solid-state switches are used in combination with step-up transformers, insulated gate bipolar transistors [174], and tirathron [175]. High-voltage pulse formation schemes are also used, in which a gas-discharge switch and a semiconductor current interrupter (SOS diodes) in an inductive drive [176, 177] are used simultaneously. The great successes achieved today in the field of developing high-power high-voltage generators based on solid-state switches give hope for the substantial successes that are possible when using them for pumping non-chain HF(DF) lasers. Study [164] reports the creation of a high-energy pulsed HF laser with a completely solid-state pump generator based on FID-switchers [178) for the first time. The electrical energy stored in the capacitors of the generator reached ~1 kJ at an open circuit voltage of 240 kV. To illustrate the prospects for the use of pump generators [178] in high-energy non-chain PP-HF (DF) lasers, this section of the thesis presents the results of research obtained by the staff of RFNC VNIIEF in conducting joint research (with the participation of the General Physics Institute of the Russian Academy of Sciences staff) HF(DF) laser with a fully solid-state pump generator [165].

5.5.1. Experimental setup

The schematic diagram of the experimental setup is shown in Fig. 5.26. The SIVD ignited between two identical duralumin electrodes at an electrode spacing of 13 cm. No special measures were taken to initiate or stabilize the discharge. The laser resonator was formed by a flat Al mirror of Al and a plane-parallel CaF_2 plate. The radiation energy was measured with a Coherent J-50 MB-HE pyroelectric measuring head. The shape of the radiation pulse was controlled by a Vigo-system Ltd. photodetector with a time resolution of ~1 ns.

The working media were mixtures of SF_6:H_2 and SF_6:D_2 gases at a pressure of $p = 0.07$–0.1 atm. Pumping the gas mixture through the DG was carried out by a fan assembly. To dispose of the spent HF (DF) molecules in the gas-dynamic path, filter cassettes were installed [165]. The laser pump generator, developed and manufactured by ZAO NPO Fid Tekhnika, consisted of three main units (Fig. 5.26): AC$\Rightarrow$DC IPP with an average power of 25 kW; control unit (CU); pulsed high-voltage pump generator, in turn, consisting of 4 generator modules (GM), which were connected in parallel to the electrodes of the discharge gap, as shown in Fig. 5.26a. Figure 5.26b shows the wiring diagram of a separate GM. The module was connected

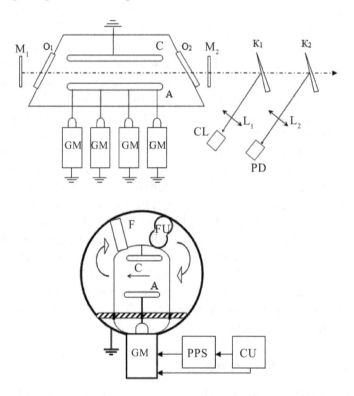

Fig. 5.26. Diagram of the experimental setup: a) optical layout; b) - connection diagram of the generator module to the electrodes. M_1, M_2 – resonator mirrors; O1, O2 – coated CaF_2 windows; K_1, K_2 – CaF_2 wedges; L_1, L_2 – lenses; CL – calorimeter; PD – photodetector; C – cathode; A – the anode; GM – generator module; F – filter; FU – fan unit; PPS – source of primary power; CU – control unit [165].

through high-voltage inputs in the dielectric cover of the discharge chamber. Inside the chamber, the connection to the anode was carried out by a series of copper conductors (using reverse current busbars) so as to ensure the minimum inductance of the leads while they are transparent to the gas flow [165].

The GMs were made according to the scheme of the 54 cascade Arkad'ev–Marx generator based on solid-state FID keys having a switching time of ≤1 ns. The temporary instability of switching on the GM with respect to the external start did not exceed 0.2 ns when the voltage at the output of the GM was changed in the range from 100 kV to 240 kV. The launch of the GM was carried out by a voltage pulse with an amplitude of 50 V and a duration of 100 ns. The total intrinsic resistance of the FID switchers at an operating voltage of 240 kV and a current of 10 kA in one GM does not exceed 1 Ohm. The switchers in this mode have a life of more than 109 pulses. The capacity at the stroke of each of the 4 GM was 7.6 nF. As in [164], the GMs were placed in metal containers filled with transformer oil with dimensions of 125 × 35 × 14 cm. Sharpening capacitors connected in parallel to the discharge gap to reduce the duration of the current, in contrast to [164], were not used in this laser variant. Reducing the number of stages in the GM made it possible to reduce the inductance of the discharge circuit so that a stable SIVD in the $SF_6:H_2$ and $SF_6:D_2$ mixtures could be realized by directly connecting the pump generator to the electrodes.

The maximum electrical energy stored in the capacitors of the pump generator was ~900 J with a maximum output voltage of 240 kV and a maximum allowable current (in short circuit mode) of 80 kA. The pump generator was designed for long-term operation at a repetition rate of discharge pulses up to 25 Hz.

During the experiments, the voltage at the discharge gap U and the discharge current I were controlled using a calibrated resistive voltage divider and a low-inductance shunt, respectively. The shunt was included in the break of the conductive earth bus.

5.5.2. Experimental results and discussion

Figure 5.27 shows a photograph of the SIVD in the discharge chamber, taken when the unit is operating in the single-pulse mode on an $SF_6:H_2$ mixture with a pressure $p = 0.09$ atm. From Fig. 5.27 it can be seen that the SIVD has a fairly homogeneous structure, with the exception of individual incomplete plasma channels, the

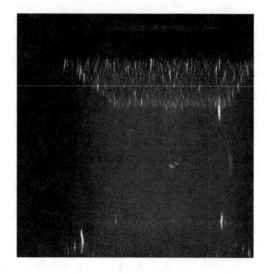

Fig 5.27. A photograph of SIVD at U_g = 240 kV and pressure of the mixture SF$_6$:H$_2$, p = 0.09 atm. Above – the cathode, below – the anode [165].

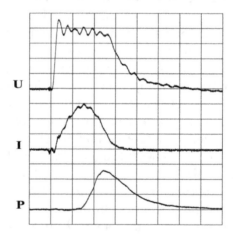

Fig. 5.28. Oscillograms of the voltage across the discharge gap U, the current through the gap I, and the lasing pulse P at U_g = 240 kV and the mixture pressure SF$_6$:H$_2$ p = 0.09 atm. U = 27 kV/div, I = 15 kA/div, P – relative units, sweep - 100 ns/div [165].

length of which does not exceed 2 cm. since the gas flow rate was sufficient for multiple gas changes in the discharge zone between pulses at a pulse repetition rate of 25 Hz.

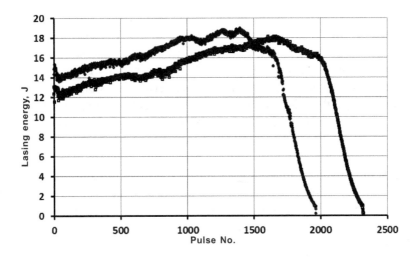

Fig. 5.29. Dependence of the laser radiation energy W on the number of the discharge pulse N in mixtures with a high content of D_2 at a pulse repetition rate of 25 Hz and a mixture pressure p = 0.1 atm. 1 – $SF_6:D_2$ = 7.5:2.5, U_g = 225 kV. 2 – $SF_6:D_2$ = 7:3, U_g = 240 kV [165].

Figure 5.28 shows the oscillograms of the voltage across the discharge gap U, the current through the gap I, and the laser generation pulse P at the no-load voltage of the GM U_g = 240 kV and the pressure of the $SF_6:H_2$ mixture, p = 0.09 atm. The duration of the discharge current at a level of 0.1 and the lasing pulse at a 0level of 0.5 amplitude values are ~290 ns and ~180 ns, respectively. At U_g = 240 kV, pressure p = 0.09 atm and optimization of the ratio of component concentrations, the specific removal of the laser energy was W_{sp} = 3.8 J/l for HF and 3.4 J/l for DF [165]. In a series of experiments at the facility described in this section, there was no system that should compensate for the loss of the mixture components dissociating in the SIVD plasma. Therefore, the laser starts (up to 90 s) were carried out in [165] on mixtures with a high content of H_2 (D_2).

Figure 5.29 shows, for illustration, the dependences of the laser radiation energy W on the number of the discharge pulse N in mixtures with an increased D_2 content at a pulse repetition rate of 25 Hz. As can be seen from Fig.5.32, the lasing energy increases from the initial value with increasing N, which is associated with the approach of the ratio of the concentrations of components to the optimal value as they dissociate. The sharp drop in energy at large

172

N in [165] was explained by the mismatch of the pump generator with the plasma load due to a decrease in the combustion voltage of the SIVD, due to a decrease in the content of SF_6 in the mixture with simultaneous accumulation of less electronegative dissociation products. However, an analysis of the experimental conditions allows us to conclude that the energy decrease was due to hydrogen burnout, since in these experiments the laser worked without changing the mixture in the chamber.

It should be noted that the modular structure of the pump generator makes it possible to increase the radiation energy of an P–P laser by simply increasing the number of connected generator modules (and, correspondingly, increasing the discharge volume). Thus, the work of a high-energy non-chain P–P HF (DF) laser with a completely solid-state pump generator was demonstrated for the first time in [165].

Conclusions to Chapter 5

The results obtained in Chapter 5 can be summarized as follows:

1. The efficiency of using SIVD for initiating non-chain reaction in SF_6:H_2 (D_2) and SF_6:C_2H_6 (C_6D_{12}) gas mixtures with the aim of creating pulsed and P–P wide-aperture HF(DF) lasers with high pulse energy and good quality is shown. The possibility of scalability of the characteristics of a non-chain electric-discharge HF(DF) laser and the possibility of creating efficient P–P lasers with high average power are demonstrated.

2. The detection of the SIVD effect allowed us to increase the generation energy of non-chain electric-discharge HF(DF) lasers by more than 40 times. A non-chain HF(DF) laser initiated by a volume discharge with an energy in a pulse of ~410 J on HF and 325 J on DF with an electrical efficiency of 4.3 and 3.4%, respectively, was created; the duration of the laser pulse in half-height was $\tau_{0.5}$~250 ns.

3. Using the example of the P–P electric-discharge HF(DF) laser with a completely solid-state pump generator based on the FID switches created at the Russian Federal Nuclear Centre, All Russian Scientific Research Institute of Experimental Physics (ILFN RFYaTs VNIIEF) [165], the possibility of improving the performance characteristics of non-chain HF(DF) lasers was demonstrated through the use of a modular design of the pump generator assembled on the basis of solid-state switchers.

Application of non-chain HF(DF) lasers

The main results presented in this chapter are published in [187, 188, 194–198, 203–217].

6.1 The main areas of application of non-chain HF(DF) lasers (a brief literature review)

In Russia today, the most high-energy and powerful non-chain P–P HF(DF) lasers have been created and are working [161, 163, 165]. For example, the average capacity of the installation described in the previous chapter is greater than 1.3 kW. Studies [76,179] describe HF(DF) lasers that operate with a repetition rate of more than 1 kHz, while they are environmentally safe and easy to use. Non-chain electrical discharge HF(DF) lasers can be operated in ordinary physical laboratories, and even in medical centres. Currently, this type of lasers is successfully used in various applications [180–188, 190–198, 203–217]. Let us give some of the most interesting, in our opinion, examples of the use of non-chain HF(DF) lasers, which are used in our laboratory.

1) Laser ignition of combustible gas mixtures. The use of non-chain HF laser to initiate the combustion of methane–air and methane–oxygen mixtures [187, 188] made it possible to investigate the dynamics of combustion in these mixtures. A number of effects were discovered, leading to a multiple increase in the burning rate. In particular, we found in [187] that the excitation of a laser spark in a closed cylindrical vessel filled with a stoichiometric methane–oxygen mixture is accompanied by an axially propagating wave of chemical reactions ('incomplete combustion wave' or 'primary wave') detected by weak radiation in the optical range. Continuously accelerating as it moves away from the spark igniter, the burning wave reaches speeds of the order of $(4–5)\cdot 10^4$ cm/s. This velocity significantly exceeds the burning wave velocity in a $CH_4:O_2$ mixture and is noticeably lower than the detonation wave velocity. After the characteristic delay times after a short laser pulse ($\tau_{las} \approx 150$ ns), reaching 500–700 μs, an almost simultaneous flash of the luminescence was observed in the entire chamber volume, accompanying the rapid ignition of the gas mixture. It should be noted that theoretical studies [189] also indicate the promise of using laser radiation with $\lambda \approx 3$ μm to accelerate and intensify the process of ignition of combustible mixtures. The search for methods to intensify the combustion of combustible gas mixtures is an urgent task in connection with the problem of creating efficient and safe detonation engines.

2) Interaction of non-chain HF laser radiation with water. Due to the fact that the absorption coefficient in water and other hydroxyl-containing liquids for HF laser radiation lines is extremely large ($\sim 10^4$ cm^{-1}), when HF laser radiation affects these liquids, it is rather easy to ensure a high rate of increase and the specific energy input to the appearance of a number of effects of both practical and general physical interest [194–197]. For example, it is possible to investigate the physics of explosive boiling up of a fluid [194, 195] and the generation of electrical signals [195, 196] or on the basis of an analysis of acoustic signals to study the properties of a substance in a supercritical state [194]. Influencing the non-chain HF laser radiation on the water surface, we investigated the nature of the appearance of an electrical signal when IR lasers interact with the water surface at radiation flux densities below plasma formation thresholds. Electric signals with amplitudes greater than 10 V were recorded and the dependences of the amplitudes of the electric signal on the laser energy were obtained [195–197]. Thresholds of volumetric explosive boiling up of water were measured under the same conditions of radiation focusing. An unambiguous connection between the effect of generating an electrical signal and the process of volumetric explosive boiling of water has been established [194]. It was found that in the case of irradiation of an open surface, an electrical signal is generated as a result of volumetric explosive boiling of water, accompanied by dropping and spraying its surface layer, destruction of the electrical layer on the surface and scattering of the electrified vapour–droplet mixture (ballo-electric effect). In the case of irradiation of the water surface covered by a transparent plate, the generation of an electrical signal occurs due to charge separation when the water surface is separated from the plate surface by a steam bubble resulting from volumetric explosive boiling up and moving the charged water surface during expansion and contraction of the steam bubble. The main advantages of using non-chain HF laser in these studies are the simplicity of the experimental setup and the ease of controlling the heating conditions. The direct effect of laser radiation on the surface of the liquid makes it possible to completely eliminate the influence on the results of studies of the vessel wall material, which, in turn, allows for a high reproducibility of the experimental results. It seems to be a very promising direction to study the possibility of treating a CVD-diamond surface with supercritical water, which is created when a powerful radiation pulse with λ ~3 μm is applied to ordinary water. It is known that diamond and sapphire [199,200] dissolve in supercritical water, while the radiation power densities of non-chain HF lasers necessary for converting water to the supercritical state are more than an order of magnitude less than those used in laser systems for treating diamond surfaces various texture patterns – diffraction gratings, etc.). The exposure to laser radiation can be carried out through the surface of the diamond plate, since they are transparent in this spectral region – in the same way as we did in [198] (closed surface mode).

In [197], the question of measuring the refractive index in a spherical wave by the Toepler method with a two-section photo detector was considered. In our experiments, we showed that when a free water surface is irradiated with a HF laser pulse in water, an elastic spherical wave is excited with an amplitude inversely proportional to the distance traveled. The minimum value of the registered refractive index gradient is ~3 $\cdot$ 10^{-8} mm^{-1}. The absolute error in measuring the refractive index is ~1.5 $\cdot$ 10^{-8}. The obtained sensitivity and accuracy of measuring the refraction index of water are superior to those previously published.

3) Optical pumping of Fe^{2+}: ZnS, Fe^{2+}: ZnSe, Fe^{2+}: CdSe, Fe^{2+}: ZnTe crystals, etc. The HF(DF) lasing spectrum falls into the absorption bands of Fe^{2+}: ZnS, Fe^{2+}: ZnSe, Fe^{2+}: CdSe, Fe^{2+}: ZnTe crystals and can be effectively converted to longer wavelengths of the average IR spectrum range of 3.5–6 μm. Therefore, in recent years, intensive studies in this direction have been conducted in Russia and China [203–215]. Such characteristics of non-chain HF(DF) lasers as

high pulsed power, the ability to generate with a high pulse repetition rate, high radiation quality and a short laser pulse allowed us to work out the technologies of producing high-quality polycrystalline Fe^{2+}: ZnS, Fe^{2+}: ZnSe samples with large transverse dimensions and obtain record characteristics of Fe^{2+}: ZnS and Fe^{2+}: ZnSe lasers [208, 209].

4) *The study of semiconductors [74]*. The possibility of obtaining high-power lasing in the spectral region of 2.6–3 μm (HF laser) and 3.5–4.1 μm (DF laser) allowed the use of an HF(DF) laser in studies of the interaction of laser radiation with semiconductor materials, which were carried out at the Prokhorov General Physics Institute of the Russian Academy of Sciences [216, 217]. In [217], we experimentally investigated the passage of radiation from a non-chain HF(DF) laser through germanium single crystals (Ge) with different thicknesses and specific electrical resistance. Based on the experimental data for the generation spectrum of the HF(DF) laser, two-photon absorption coefficients in Ge are calculated - $K_2 = 55\pm10$ cm/GW for $\lambda = 2.8$ μm. The results are in good agreement with theory. In [216], we developed a numerical model that describes the experimental results with good accuracy and allows us to consider the dynamics of the process at any given time during a laser pulse. It was shown that for high-power laser radiation with a wavelength of $\lambda = 2.5$–4 μm, thin Ge coatings can effectively equalize the energy distribution over the beam aperture. The influence of the parameters of the laser pulse and the energy distribution over the beam aperture on the process of radiation passing through a germanium crystal is investigated. The obtained data can be used to create effective limiters for the power of laser radiation in the 2.5–4 μm range.

The next section presents the most interesting applications of non-chain HF(DF) lasers for creating 3.7–5 μm laser emitters with high pulse energy that are tunable in the spectral region.

6.2. Possibilities for expanding the lasing spectrum of an HF(DF) laser

For various practical applications of HF(DF) lasers, in addition to improving the energy characteristics, urgent tasks are to find ways to efficiently control the lasing spectrum of the laser and efficient methods to expand its lasing spectrum to the long-wave region of the spectrum where there are no high-power lasers. In addition to searching for efficient frequency converters for non-chain HF(DF) lasers in the long-wavelength region of the spectrum, there is a possibility of increasing the fraction of the energy of non-chain HF(DF) lasers emitted at lower frequencies [3]. Indeed, due to the anharmonicity of the HF(DF) molecule, the energy of photons emitted during the transition from neighbouring vibrational–rotational levels decreases with increasing number of the HF (n) level. Those. when switching from high levels, the radiation wavelength will be longer. In a typical non-chained HF(DF) laser, almost all energy is concentrated in the spectral region of 2.7–3 μm (HF laser) and 3.6–4.1 μm (DF laser) [19]. These are, as a rule, cascade transitions from the first 3-4 oscillatory levels. The possibility of controlling the laser spectrum in this range is shown in [16]. To obtain longer wavelength radiation, it is necessary that molecules excited to a level of 7 or higher should occur in the reactions. In [17, 18, 35], it was also pointed out that the spectral range of HF(DF) laser generation was increased with variations in the composition of the working medium. If molecules with low dissociation energy, for example, HI, are used as hydrogen carriers (RH), then lines with a long wavelength corresponding to transitions from high vibrational levels (up to $n = 9$) were also observed in the spectrum of non-chain HF laser [218, 219] . In experiments with a DF laser, wavelengths in the range $\lambda = 4.74 \div 4.88$ were observed, which corresponded to vibrational-rotational transitions with $n = 9$ vibrational levels at $n = 8$ [3]. However, note that one should not expect that in this way a significant fraction of the energy in the long-wavelength region can be obtained (with $\lambda > 4.5$ μm). As noted above, the generation in the HF laser is of a cascade nature. Therefore, even if we consider the idealized variant, when as a result of a non-chain reaction we received all DF molecules excited to the level $n = 8$ (n is the number of the vibrational level), then the radiation emitted by the transition from the 8th to the 7th level will be even in the ideal case, a value less than 1/8 of the total pulse energy. Taking into account the fact that the stability of SSVD in SF_6:DI mixtures is significantly lower than in pure SF_6, we have to admit that this method of spreading the spectrum can be only of theoretical interest.

The possibility of increasing the generation energy of a HF laser at wavelengths with $\lambda > 14$ μm (generation at rotational transitions) was theoretically shown in [83]. The possibility of lasing on rotational transitions in the far infrared spectral region was experimentally demonstrated in Refs. [84, 85], but most of these studies were carried out with chain lasers [86]. Apparently, it is possible to achieve progress in this area of research with the use of non-chain electric-discharge lasers, but today there is no urgent need for this. The most promising, for practical implementation, at present, is the use of radiation from non-chain HF(DF) lasers for optical pumping of the crystal structures of chalcogenides doped with transition metal ions, for example, Fe^{2+}: ZnS, Fe^{2+}: ZnSe, Fe^{2+}: ZnTe, etc., in order to obtain a powerful pulsed and and P–P lasing in the spectral range of 4–6 µm. Until 2014, there was no significant success in this area, despite the fact that technologies were developed for producing new types of high-quality crystal structures suitable for converting HF(DF) laser radiation. For example, modern technologies allow creating large crystals of chalcogenides with apertures of more than 4 cm, doped with transition metals (Fe^{2+}:ZnS, Fe^{2+}:ZnSe and Fe^{2+}:ZnTe), on which lasing can be obtained in the spectral region of 4–6 µm [87–91]. Let us consider active media based on chalcogenide crystal structures of Fe^{2+}:ZnSe and Fe^{2+}:ZnS in more detail.

6.2.1 Chalcogenide crystals (ZnSe, ZnS) doped with transition metal ions (Fe^{2+})

Optical pumping of Fe^{2+}:ZnSe and Fe^{2+}:ZnS crystals was always carried out using Er:YAG or Er:Cr:YSGG solid-state lasers. Note the parameters of these lasers are significantly inferior to non-chain chemical HF(DF) lasers [89]. It is noteworthy that the lines of non-chain HF(DF) laser, initiated by the SSVD, fall in the absorption region of these crystals and can be used for their effective optical pumping.

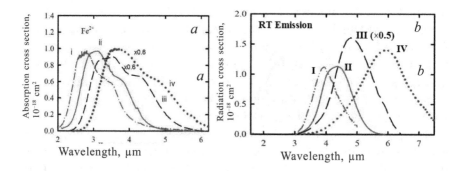

Fig. 6.1. Spectral dependences: a) absorption cross sections of the Fe^{2+} ion in the ZnS (i), ZnSe (ii), CdSe (iii) and CdTe (iv) matrix at room temperature; b) the luminescence cross section of the Fe^{2+} ion in the ZnS (iv), ZnSe (v), CdSe (vi) and CdTe (vii) matrix, at room temperature [91].

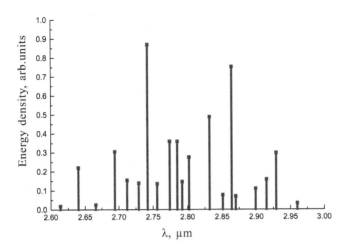

Fig. 6.2. Relative intensity of the lasing lines of a pulsed non-chain HF laser used in experiments on optical pumping of Fe:ZnSe and Fe: ZnS crystals.

Figure 6.1 shows the spectral dependences of the absorption and luminescence cross sections for the Fe^{2+} ion in various crystal structures at room temperature [91]. Figure 6.2 shows a typical generation spectrum of a non-chain HF laser initiated by OCP, obtained at the facility described in the previous section. From Fig.6.1a and Fig.6.2 it is clearly seen that the spectrum of non-chain electric-discharge HF laser falls entirely into the absorption band of iron ion Fe^{2+} in crystalline matrices – ZnS, ZnSe and CdSe, and the spectrum of DF laser (Fig. 1.5) is well absorbed in crystal structures ZnSe: Fe^{2+}, CdSe: Fe^{2+} and CdTe: Fe^{2+}.

Figure 6.16 shows the dependences of the decay time of the luminescence of the upper level of the Fe (Cr) ions in the ZnSe and ZnS matrices on temperature taken from [91]. It can be seen from this figure that, at room temperature, the lifetime of the upper levelin the Fe: ZnSe crystal structure is $\tau_{life} \sim 360$ ns. Note that the typical duration of a laser pulse in non-chain HF(DF) lasers, even in a HF laser, with an output energy of ~100 J, does not exceed 350 ns. Therefore, it seems very promising to use for optical pumping of a Fe:ZnSe crystal radiation of non-chain electric-discharge HF laser to obtain lasing in the spectral region of 4–5 μm at room temperature of the active element. It should be noted that, in addition to high pulsed power and energy, non-chain HF(DF) lasers initiated by SIVD have another important advantage over solid-state Er:YAG or Er:Cr:YSGG

180

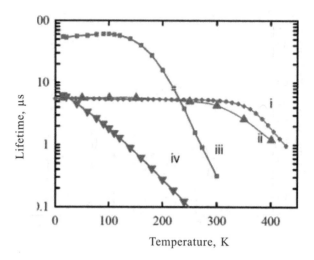

Fig. 6.3. The dependence of the lifetime of the upper level on the temperature in the crystal structures: (i) – Cr:ZnSe, (ii) – Cr:ZnS; (iii) – Fe: ZnSe, (iv) – Fe:ZnS [91].

solid-state lasers — this is the ability to work with large apertures and high quality laser beam. This circumstance is extremely useful in developing the technology of creating laser crystals. Indeed, as experience shows when creating new crystalline laser media, crystal manufacturers cannot immediately achieve uniformity of crystal characteristics throughout the entire volume. Therefore, when the pump beam is small, it is necessary to scan the entire surface of the crystal with a small spot. In this case, it becomes very difficult to interpret the results of laser experiments. Still more aggravated when the crystal has to be cooled to cryogenic temperatures. In this case, the possibility of prompt adjustment of the experiment scheme is completely lost. When the crystal is damaged, it is necessary to perform many additional operations on installation, adjustment, etc. As a result, the simplest experiment comparing the laser characteristics of two samples requires considerable time. It is quite another thing when you can work with a crystal at room temperature, without using special crystal cooling devices and use laser beams with a size of more than 3 mm for pumping. Due to the large beam size, the parameters of the crystal structure are averaged over the pump spot aperture already in one experiment. Therefore, when it is necessary to quickly compare several samples of crystal structures under the same pumping conditions, the use of non-chain HF(DF) lasers is more preferable. Below are the results of studies

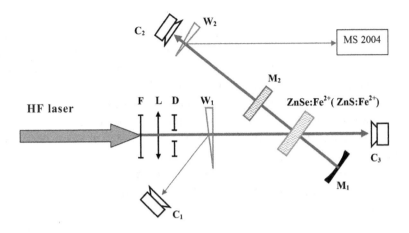

Fig. 6.4. Scheme of the experiment: M_1, M_2 – resonator mirrors; F – a set of calibrated attenuators; L – spherical lens; D – iris diaphragm; W1, W2 – optical wedges from CaF_2; C_1, C_2, C_3 – calorimeters, MS 2004 - monochromator from SOL instruments.

of the characteristics of $ZnSe:Fe^{2+}$ and $ZnS:Fe^{2+}$ lasers excited by non-chain HF lasers.

6.2.2 Experimental setup and measurement technique

Figure 6.4 shows the experimental setup. The technique for studying the characteristics of lasers on $ZnSe:Fe^{2+}$ samples ($ZnS:Fe^{2+}$) is described in detail in [203–213]. All studies were conducted at room temperature of the active elements. The radiation of a non-chain electric-discharge HF laser with a half-amplitude pulse $\tau_{las} \sim$ 140 ns, weakened by a set of calibrated light filters F, was focused on the sample surface with a spherical lens L with a focal length of 45 cm. In these experiments, the range was $d = 3.9 \div 14.5$ mm using an iris diaphragm D. A non-chain HF laser operated in a single-pulse mode, its characteristics were described in section 5.3.1. Under the conditions of these experiments, the crystal is pumped on a set of lines in the spectral range $\lambda = 2.7 \div 3.1$ μm (see Fig. 6.2) generated in a single pulse. Figure 6.2 shows the lasing spectrum of the non-chain laser that was used in this experiment. The angle of incidence of the pump radiation on the crystal surface was $\sim 20°$. The crystals were installed in the resonator so that their polished faces were perpendicular to the optical axis of the resonator. The values of the HF laser radiation incident on the sample, transmitted through the sample, and the lasing of solid-state lasers were measured with C_1 and C_3 calorimeters manufactured by Gentec-EO and Molectron (C_2), respectively. The ZnSe:Fe^{2+} laser resonator with a length of $l_r = 130 \div 180$ mm was formed by a concave mirror M_1 (a mirror with a gold coating on a quartz substrate) with a radius of curvature of 0.5 m and a flat output mirror M_2. The output mirror M_2 was made from plane-parallel CaF$_2$ plates with an interference coating, having a reflection coefficient at wavelengths $\lambda = 4$–5 μm, $R = 40\%$, $R = 60\%$ and $R = 80\%$, in some experiments with the output mirror of the ZnSe:Fe^{2+} laser was an Si plate (reflectance $R \approx 50\%$).

For each sample of the crystal, the coupling of the resonator was optimized and detailed studies of the dependence of the output ZnSe:Fe^{2+} energy (ZnS:Fe^{2+}) of the laser on the pump energy were carried out with an output mirror, on which the lasing efficiency reached its maximum. Initially, studies were carried out with an average irradiation spot diameter on a 3.9 mm sample, then gradually the spot diameter increased until the generation efficiency of the ZnSe:Fe^{2+} (ZnS:Fe^{2+}) laser began to decrease due to the development of parasitic generation [204], which is characteristic of disk lasers [220].

The lasing spectra of ZnS:Fe^{2+} and ZnSe:Fe^{2+} lasers were measured similarly to [206] using an MS 2004 monochromator (SOL instruments) with diffraction gratings of 75, 150 and 300 lines/mm. At the output of the monochromator there was an array of pyroelectric photodetectors APD brand HPL-256-500 (HEIMANN Sensor). The line contained 256 pyroelectric elements with a receiving site width of 50 μm, the length of its receiving part was 12.8 mm [205].

6.2.3 Samples of crystal structures used in experiments

In the experiments at the facility described in the previous section, the most diverse samples of ZnSe:Fe^{2+} crystal structures (ZnS:Fe^{2+}) were investigated: single-crystal doped during growth [210, 211]; single crystal diffusion doped from the surface [26]; A wide class of polycrystalline active elements also diffusion doped from the surface, which were created at the Institute of Chemistry of High-Purity Substances of the Russian Academy of Sciences taking into account the results of research conducted in our laboratory. Among the active elements of ZnSe:Fe^{2+} (ZnS:Fe^{2+}) lasers created at the Institute of Chemistry of High-Purity Substances of the

RAS in the laboratory of E.M. Gavrishchuk were polycrystalline
doping structures, which were carried out by long-term annealing
in ampoules [203–206]; samples subjected to long-term gas-static
treatment [208, 209, 213]; ceramic samples prepared by hot pressing;
as well as specially prepared structures in which the doping profile
had a maximum inside the crystal [204, 207], etc. A detailed
analysis of the characteristics of these samples and their preparation
technology are contained in [212, 214].

6.2.4 Experimental results and discussion

Figure 6.5 shows typical (for the polycrystalline samples with
diffusion doping from the surface) dependences of the transmittance
T on the density of the HF laser radiation energy incident on the
surface W_{in} taken at different spot sizes of pump radiation $a \times b$ (a
is the size of the major axis of the ellipse) on the surface ZnSe:
Fe^{2+} crystal [204]. These dependences were obtained in the absence
of lasing of the ZnSe:Fe^{2+} laser (for this, the resonator mirrors were
closed with a scattering screen). As can be seen from Fig. 6.5, the
nature of the transmission strongly depends on the size of the spot. At
relatively small spot sizes, the transmission monotonically increases
with increasing pump energy density, experiencing a weak tendency
to saturation (the so-called bleaching of the sample is observed).

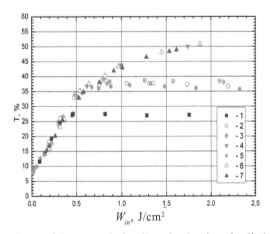

Fig. 6.5. Dependence of the transmission T on the density of radiation energy incident
on the surface of the HF laser W_{in} in the absence of lasing at different pump spot
sizes: 1 – $a \times b = 7.6 \times 8.1$ mm; 2 – $a \times b = 6.6 \times 7$ mm; 3 – $a \times b = 6.3 \times 6.8$
mm; 4 – $a \times b = 5.3 \times 5.4$ mm; 5 – $a \times b = 6.3 \times 6.6$ mm; 6 – $a \times b = 4 \times 4.1$
mm; 7 – $a \times b = 2.6 \times 2.8$ mm [204].

184

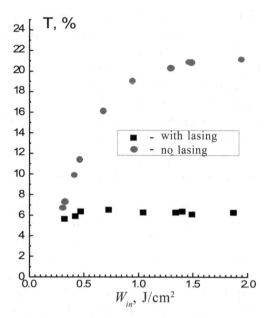

Fig. 6.6. Dependence of the transmission T on the density of laser radiation energy incident on the surface of the HF laser W_{in} in the absence of lasing and under the conditions when the mirrors of the ZnSe:Fe^{2+} laser are open.

Increasing the pumping spot results in saturation of the transmission at lower pump energy densities. This behaviour of T versus W_{in} is due to the development of parasitic lasing (superluminescence) with large pumping spot sizes [204, 209] in the surface layer of the sample, where the maximum gain of the active medium is reached, because the concentration of the doping component is maximum in this layer. With open mirrors, lasing along the optical axis of the resonator develops faster than spurious lasing in the surface layer of the sample until the maximum spot size reaches a certain critical value [204, 209].

It should be noted that in the presence of lasing, the transmittance of the sample under study changes noticeably. Figure 6.6 shows the transmission dependences of the polycrystalline ZnSe:Fe^{2+} structure under the conditions when there is lasing and when not (laser mirrors are closed). In this case, the increase in the pump radiation absorption in the crystal structure is associated with an increase in the lasing radiation density in the same polycrystal region (a positive feedback occurs). This circumstance must be taken into account in order to avoid radiation damage to the crystal structure.

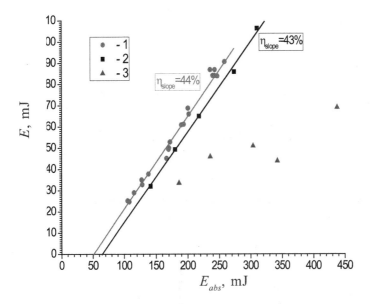

Fig. 6.7. The dependences of the lasing energy of a ZnSe:Fe^{2+} laser on the HF laser energy absorbed in the crystal structure for different pump spot diameters: 1 – d = 3.9 mm; 2 – d = 4.5 mm; 3 – d = 5.3 mm.

In the sample, in which the doped layer is thin (the sample was annealed in the ampoule for only 3 days), the generation failure with increasing pumping spot occurs much earlier than on the samples, where iron ions were diffused over long distances. Figure 6.7 shows the dependences of the output energy of a ZnSe:Fe^{2+} laser on the HF laser energy absorbed in the sample, with different pump spot diameters. From this figure it can be seen that the generation failure occurs at an irradiation spot diameter of only 5.3 mm.

Figure 6.8a shows oscillograms of lasing of ZnS:Fe^{2+} and HF lasers, and Fig. 6.8b shows for comparison an oscillogram of lasing of ZnSe:Fe^{2+} laser pulses taken at the same spot diameter and close to the absorbed energy. The ratio of the amplitudes of the signals in Fig. 6.8 does not reflect the real ratio of the laser powers, the amplitudes of the signals were regulated by light filters for convenience of perception. As can be seen from this figure, the ZnS:Fe^{2+} laser lasing pulse has a longer leading front and a noticeably shorter duration than the ZnSe:Fe^{2+} laser lasing pulse. It also does not have a pronounced short and powerful front peak characteristic of a ZnSe: Fe^{2+} laser pulse for large values of the pump energy density.

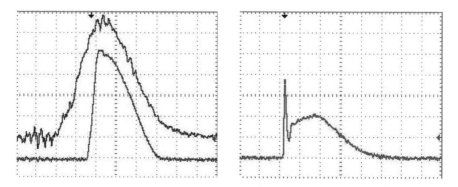

Fig. 6.8. Waveforms of generation pulses: a) ZnS:Fe^{2+} (lower beam) and HF (upper beam) lasers; b) ZnSe:Fe^{2+} laser. Sweep 50 ns / div.

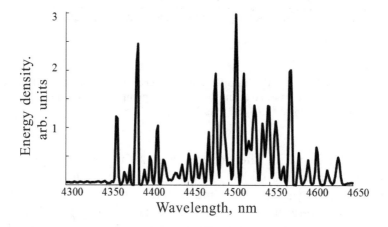

Fig. 6.9. ZnSe:Fe^{2+} laser spectrum when pumping a polycrystalline structure with a non-chain HF laser. Grating 150 lines/mm.

Figures 6.9 and 6.10 show typical lasing spectra of ZnSe:Fe^{2+} and ZnS:Fe^{2+} lasers, which were observed under non-selective resonator conditions (see Fig. 6.1). The distribution of peak power and wavelength energy varies from pulse to pulse at a constant total lasing energy. The spectra show a ruled structure, which is due to the complex dynamics of the formation of the radiation field in the resonator [221]. Earlier, in special experiments, we established that this structure of the spectrum was not related to the presence of thin plates that could play the role of a Fabry–Perot filter, or gaseous impurities, which under the experimental conditions could impose a linear structure of the generation spectrum [213, 206]. A similar

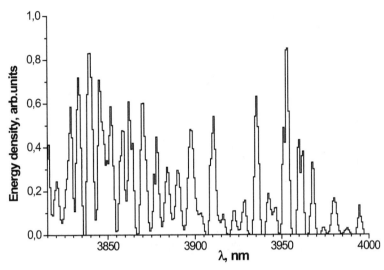

Fig. 6.10. ZnS:Fe^{2+} laser spectrum when pumping a polycrystalline structure with a non-chain HF laser. Grating 300 lines/mm.

conclusion about the nature of the line spectrum of generation on polycrystalline chalcogenides doped with iron and chromium ions was also made by other research groups [222, 223].

It should be noted that the dependences shown above (Figs. 6.5-6.7) were obtained even on the first samples of polycrystals of 20 mm diameter, which were doped in quartz ampoules, by prolonged annealing (several days) in argon atmosphere [203]. Studies have shown [209] that increasing the size of polycrystals and gas-static treatment (at pressures of 1000 atm.) at high temperatures (T = 1250°C) in graphite containers can significantly improve the laser characteristics of the obtained polycrystalline ZnSe:Fe^{2+} structures. Further improvement in the technology of creating such polycrystalline structures made it possible to achieve record-breaking parameters of ZnSe:Fe^{2+} (ZnSe:Fe^{2+}) lasers.

Figures 6.11 and 6.12 show the dependences of the output energy of the Fe^{2+}:ZnSe and Fe^{2+}:ZnS lasers on the absorbed energy of the non-chain HF laser. As can be seen from these graphs, despite the fact that the experiments were carried out at room temperature, the conversion efficiency of the non-chain HF laser radiation in polycrystalline samples is very high, the following results are currently obtained: Fe:ZnSe laser – energy 1.53 J; differential efficiency 53%; total efficiency in terms of absorbed energy 48%

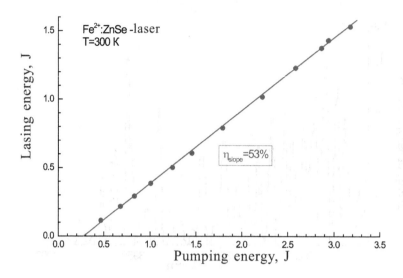

Fig. 6.11. The dependence of the lasing energy of a ZnSe:Fe^{2+} laser on the absorbed energy of an HF laser [209].

(in the spectral range 4.3–4.8 μm); Fe:ZnS laser – energy 660 mJ; differential efficiency 36%; total efficiency in terms of absorbed energy 26% (in the spectral range 3.7–4.0 μm).

In [213], a powerful P–P Fe^{2+}:ZnSe laser excited by non-chain HF laser radiation was used, which generated an output energy of 1.67 J with an efficiency of ~43% absorbed in the polycrystal and operating in the P–P mode without using special polycrystal sample cooling systems. With a pulse repetition rate of 20 Hz, an average laser power of ~20 W with energy in a single pulse of ~1 J and full efficiency in terms of absorbed power of ~40% was obtained [213].

6.2.5 Analysis of results

The results of the studies presented in this section, as well as the work of other research groups [203–215], show that the non-chain HF(DF) laser initiated by the SIVD (SSVD) is a versatile and extremely convenient source of pumping for a whole class of chalcogenide crystals doped with iron ions (modifications are possible with the additional salt alloying by Mn ions). The use of crystal matrices based on ZnS, ZnSe, CdSe and CdTe, as well as their modifications doped with iron ions, will allow you to create laser

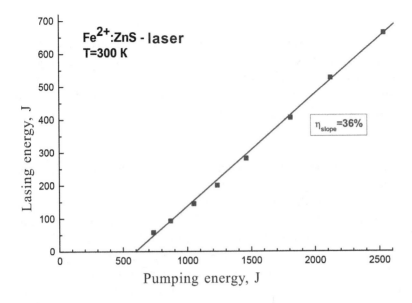

Fig. 6.12. Dependence of the lasing energy of a ZnS:Fe^{2+} laser on the absorbed energy of an HF laser [208].

systems that allow you to get powerful laser radiation tunable in the spectral range of 4–6 µm. It is noteworthy that in these systems one electric-discharge HF(DF) laser will be used as the pump source. An important advantage of this pump source is the ability to pump crystals at room temperature of the active element. (To increase the efficiency, not deep cooling of the crystal should be carried out).

Conclusions to chapter 6

It is shown that the radiation of a non-chain HF laser allows efficient pumping of ZnSe and ZnS crystals doped with iron ions without using cryogenic systems for cooling crystals. At room temperature, the conversion efficiency of non-chain HF laser radiation in a Fe^{2+}:ZnSe polycrystal can exceed 48%, i.e. achieves the maximum possible efficiency when a crystal is excited by a 3 µm laser source. Based on studies of the characteristics of laser systems based on Fe^{2+}:ZnSe and Fe^{2+}:ZnS excited by the non-chain HF laser, physical principles were proposed for developing the technology for producing high-quality chalcogenide active elements with large transverse dimensions, including the spectral range 4–5 µm) based on ZnSe and ZnS polycrystals doped with iron ions.

At room temperature of the polycrystal, the following were obtained: Fe^{2+}:ZnSe laser – energy 1.53 J; differential efficiency 53%; total efficiency in terms of absorbed energy 48% (in the spectral range of 4.3–4.8 µm); Fe^{2+}:ZnS laser – energy 660 mJ; differential efficiency 36%; the total efficiency in terms of absorbed energy is 26% (in the spectral range 3.7–4.0 µm).

Conclusion

As a result of the research conducted in the book, the following results can be distinguished. First:

1. A new form of discharge has been discovered – a self-initiated volume discharge (SIVD), which is a form of self-sustained volume discharge (SSVD) obtained without preionization in gas mixtures containing highly electronegative polyatomic gases SF_6, C_3F_8, C_2HCl_3 and C_3H_7I. The SIVD extends across the gap perpendicular to the direction of the applied electric field through the formation of diffuse plasma channels. The formation of the SIVD is achieved by creating small-scale inhomogeneities on the cathode surface, which significantly enhance the local electric field, and/or UV illumination of the cathode. Uniform in volume of the discharge gap (DG), the energy input into the SIVD plasma is realized through non-stationary self-organizing processes of determining the distribution of current density in diffuse channels.

2. It is shown that the effect of the implementation of SIVD is associated with the existence of mechanisms for limiting the current density in the diffuse channel. It has been established that in gas mixtures of a non-chain HF(DF) laser, the main physical processes causing this mechanism are the dissociation of SF_6 and other molecular components of the mixture by electron impact, as well as electron–ion recombination.

3. The electrical characteristics of a single plasma formation (diffuse channel) were studied to quantitatively measure the effect of limiting the current density. Criteria for the qualitative and quantitative assessment of the effect of various gas additives on the stability of SIVD in SF_6-based gas mixtures are established.

4. The electrophysical properties of SF_6 and its mixtures at high temperatures were determined on the basis of an analysis of the current and voltage waveforms of the SIVD ignited in a gas previously irradiated by a CO_2 laser. A significant increase was found in the discharge voltage with increasing temperatures.

It is shown that this effect is due to the attachment of electrons to vibrationally excited molecules. The temperature dependence of the reduced electric field strength $(E/N)_{cr}$ in SF_6 was obtained in the temperature range $T = 300–2300$ K.

5. It has been established that heating of SF_6 leads to discharge filamentation and the development of plasma instabilities in the volume of the discharge gap, and not from the surface of the electrodes. It is shown that the most probable mechanism of discharge contraction in working mixtures of a non-chain HF(DF) laser is a decrease in the rate of electron–ion recombination, which ceases to compensate for the increase in electron concentration due to the processes of their detachment from negative ions.

6. It is shown that the implementation of the SIVD allows one to significantly increase the aperture and discharge volume of non-chain HF(DF) lasers, to increase their reliability and service life. Pulsed lasers initiated by the SIVD with an electrode spacing of 27 cm, a volume of up to 70 liters and an emission energy of 410 J per HF and 330 J per DF with an efficiency of 4.3% and 3.4%, respectively, were constructed as well as non-chain repetitively pulsed HF lasers with an average power over 1 kW.

7. High-energy laser systems based on optically excited using a non-chain HF(DF) laser crystalline structures of $ZnSe:Fe^{2+}$ and $ZnS:Fe^{2+}$ and emitting at room temperature in the spectral range of 3.7–5 μm have been proposed and implemented. An energy of 1.53 J was obtained on the polycrystalline structure of $ZnSe:Fe^{2+}$ in the spectral range 4.3–4.8 μm with an efficiency of 48% for the absorbed energy. An energy of 660 mJ was obtained in the $ZnS:Fe^{2+}$ polycrystalline structure in the spectral range 3.7–4 μm with an efficiency of 26% for the absorbed energy.

Bibliography

1. Gross R. W. F., Bott J. F. Handbook of chemical lasers, New York, Wiley-Interscience, 1976. 754 p. - 1976.
2. Handbook of Lasers / Ed. A.M. Prokhorov. - Moscow, Sov. radio, 1978.
3. Bashkin A.S., Igoshin V.I., Oraevsky A.N., Scheglov V.A. Chemical lasers. - M.: Science, 1982.
4. Apollonov V.V. High power lasers in our life. - Nova, 2016.
5. Lillesand T., Kiefer R. W., Chipman J. Remote sensing and image interpretation. - John Wiley & Sons, 2014.
6. Cook J. High-energy laser weapons since the early 1960s, Optical Engineering. - 2012. - V. 52. - No. 2. - P. 021007.
7. Perram G.P., Marciniak M.A. and Goda M. High energy laser weapons: technology overview, Proc. SPIE. - 2004. - V. 5414. - P. 1–25.
8. Boreisho A.S. Powerful mobile chemical lasers, Kvant. elektronika. - 2005. - V. 35. - No. 5. - P. 393-406.
9. Ablekov V.K., Denisov Yu.N., Proshkin V.V. Chemical lasers. - Moscow, Atomizdat, 1980.
10. Deutsch T.F. Molecular laser action in hydrogen and deuterium halides, Applied Physics Letters. - 1967. - V. 10. - P. 234.
11. Kompa K.L. and Pimentel G.C. Hydrofluoric Acid Chemical Laser, The Journal of Chemical Physics. - 1967. - V. 47. - P. 857.
12. Yeletsky A.V. Processes in chemical lasers, Usp. Fiz. Nauk - 1981. - V. 134. - No. 2. - P. 237-278.
13. Bortnik I.M. Physical properties and dielectric strength of SF6. - Moscow, Energoatomizdat, 1998.
14. Baranov V.Yu., Vysikaylo F.I., Demyanov A.V., Malyuta D.D., Tolstov V.F. Parametric studies of a pulsed non-chain HF laser, Kvant. elektronika. - 1984. - V. 11. - No. 6. - P. 1173-1178.
15. Agroskin V.Y., Bravy B.G., Chernyshev Y.A., Kashtanov S.A., Kirianov V.I., Makarov E.F., Papin V.G., Sotnichenko S.A., Vasiliev G.K. Aerosol sounding with a lidar system based on a DF-laser, Appl. Phys. B. - 2005. - V. 81. - No. 8. - P. 1149–1154.
16. Gordon E.B., Matyushenko V.I., Nadhin A.I., Sizov V.D., Sulemenkov I.V. Control of the spectral composition of radiation from a high-power pulsed chemical HF laser, Khim. Fizika. - 1993. V. 12. - No. 10. - P. 1359.
17. Obara M., Fujioka T. Pulsed HF Chemical lasers from Reactions of Fluorine Atoms with Benzene, Toluene, Xylene, Methanol, and Acetone, Jap. J. Appl. Phys. - 1975. - V. 14. - No. 8. - P. 1183-1187.
18. Green W.H., Lin M.C. Pulsed Discharge Initiated Chemical Lasers. III. Complete Population Inversion in HF, The Journal of Chemical Physics. - 1971. - V. 54. - P. 3222.
19. Lomaev M.I., Panchenko A.N., Panchenko N.A. Spectral characteristics of radiation of nonchain HF (DF) lasers pumped by a volume discharge, Optika atmosfery i

okeana. - 2014. - V. 27. - No. 4. - S. 341-345.

20. Kazantsev S.Yu., Wide-aperture HF (DF) lasers initiated by self-sustained volume discharge. Candidate dissertation. Phys.-Math. sciences. 04/01/21 / Moscow, 2002 .-- 150 s.

21. Bagratashvili V.N., Knyazev I.N., Kudryavtsev Yu.A., Letokhov V.S. High-pressure electrochemical HF laser, Pisma v Zh. Eksper. Teor. Fiz. - 1973. - V. 18. - No. 2. - S. 110-113.

22. Papagiakoumou B., Papadopoulos D.N., Makropoulou M., Khabbaz M. G., Serafetinides A.A. Pulsed HF laser ablation of dentin, Proc. SPIE. - 2005. - V. 5777. - P. 978-981.

23. Izatt J., Sankey N., Partovi F. Ablation of calcified biological tissue using pulsed HF laser radiation, IEEE J. Quanum Electronics. - 1990. - V. 26. - P. 2261-2269.

24. Furzikov N.P. The nature of corneal and skin ablation by IR laser radiation, Kvant. Elektronika. - 1991. - V.18. - No. 2. - P. 250–253.

25. Wolbarsht M.L. Laser Surgery: CO_2 or HF, IEEE J. Quanum Electronics. - 1984. - V. 20. - P. 1427.

26. Velikanov S.D., Danilov V.P., Zakharov N.G., Ilyichev N.N., Kazantsev S.Yu., Kalinushkin V.P., Kononov I.G., Nasibov A.S., Studenikin M.I., Pashinin P.P., Firsov K.N., Shapkin P.V., Schurov V.V. ZnSe: Fe^{2+} crystal laser pumped by radiation from a nonchain electric-discharge HF laser at room temperature, Kvant. Elektronika.. - 2014. - V. 44. - No. 2. - P. 141-144.

27. Baranov V.Yu., Vysikaylo F.I., Demyanov A.V., Kochetov I.V., Malyuta D.D., Tolstov V.F. Investigation of the spectral time and energy characteristics of a pulsed chemical non-chain HF laser. Moscow, TSNIIATOMINFORM, preprint of the IAE. - 1983 .-- 3780/14, 31 s.

28. Jones C.R., Buchwald M.I. Ammonia laser optically pumped with an HF laser, Optics Communications. - 1978. - V. 24. - P. 27.

29. Evans D.K., Robert D. McAlpine, McClusky F.K. Laser isotope separation and the multiphoton dissociation of formic acid using a pulsed HF laser, Chemical Physics. - 1978. - V. 32. - P. 81-91.

30. Fedotov O.G., Fomin V.M. An electric discharge chemical HF laser is a highly coherent source for IR holography. Prospects for practical application, Journal of Technical Physics. - 2018. - V. 88. - No. 2. - P. 258-264.

31. Lazarenko V.I., Velikanov S.D., Pegoev I.N. Sinkov S.N. and Frolov Yu.N. Analysis of DF laser applicability to SO_2 remote sensing in the atmosphere, Proc. SPIE. - 2001. - V. 4168. - P. 232–235.

32. Velikanov S.D., Elutin A.S., Kudryashov E.A., Pegoev I.N., Sinkov S.N., Frolov Yu.N. The use of a DF laser for the analysis of hydrocarbons in the atmosphere, Kvant. Elektronika.. - 1997. - V. 24. - No. 3. - S. 279–282.

33. Jensen R.J., Rice W.W. Electric discharge initiated SF_6-H2 and SF_6-HBr chemical lasers, Chemical Physics Letters. - 1970. - V. 7. - P. 627-629.

34. Jacobson T.V., Kimbell G.H. Transversely pulse-initiated chemical lasers: Atmospheric-pressure operation of an HF laser, Journal of Applied Physics. - 1971. - V. 42. - P. 3402-3405.

35. Jacobson T.V., Kimbell G.H. WC 8 Parametric studies of pulsed HF lasers using transverse excitation, IEEE J. Quantum Electronics. - 1973. - V. 9. - P. 173-181.

36. Zapolsky A.F., Yushko K.B. An electric discharge laser based on an SF6 - H2 mixture pumped from an inductive storage device, Kvant. Elektronika. - 1979. - V. 6. - P. 408.

37. Wlodarczyk G. A Photopreionized Atmospheric Pressure HF Laser, IEEE J. Quantum Electronics. - 1978. - V. 14. - P. 768-771.

38. Midorikava M., Sumida S., Sato Y., Obara M., Fujioka T. An UV preionised self sustained discharge HF / DF laser, IEEE J. Quantum Electronics. - 1979. - V. 15. - P. 190.

39. Anderson N., Bearpark T., Scott S.J. An X-ray preionised self sustsined discharge HF / DF laser, Applied Physics B. - 1996. - V. 63. - P. 565-573.

40. Voigner F., Gastaund M. Improved performance of a double discharge initiated pulsed HF chemical laser, Applied Physics Letters. - 1974. - V. 25. - P. 649.

41. Pummer H., Breitfeld W., Welder H., Klement G., and Kompa K.L., Parameter study of 10-J hydrogen fluoride laser, Applied Physics Letters. - 1973. - V. 22. - No. 7. - P. 319.

42. Puech V., Prigent P., Brunet H., High-Efficiency, High-Energy Performance of a Pulsed HF Laser Pumped by Phototriggered Discharge, Applied Physics B. - 1992. - V. 55. - P. 183–185.

43. McDaniel I., Nigen W. Gas lasers. - Mosocw, Mir, 1986 .-- 548 p.

44. Korolev Yu.D., Mesyats G.A. Physics of pulse breakdown of gases. - Moscow, Nauka, 1991 .-- 224 p.

45. Reiser Yu.P. Physics of gas discharge. - Moscow, Nauka, 1991 .-- 536 p.

46. Bychkov Yu.I., Korolev Yu.D., Mesyats G.A. Pulsed discharge in a gas under conditions of intense electron ionization, Usp. Fiz. Nauk. - 1978. - V. 126. - No. 3. - S. 451–475.

47. Mesyats G.A., Korolev Yu.D. Volume discharges of high pressure in gas lasers, Usp. Fiz. Nauk. - 1986. - V. 148. - No. 1. - P. 101–122.

48. Osipov V.V. Independent volume discharge, Usp. Fiz. Nauk.. - 2000. - V. 170. - No. 3. - P. 225-245.

49. Slovetsky D.I., Deryugin A.A. The energy distribution functions of electrons and the interaction of electrons with polyatomic fluorine-containing gases: In the book. Chemistry of plasma. Issue 13. Ed. B.M.Smirnov - Moscow, Energoatomizdat. - 1987. - P. 240–277.

50. Gordon E. B., Matyushenko V. I., Repin P. B., Sizov V. D. The energy price of the formation of fluorine atoms in a pulsed electric discharge, Khim. Fizika. - 1989. - V. 8. - No. 9. - P. 1212.

51. Bashkin A.S., Oraevsky A.N., Tomashev V.N., Yuryshev N.N. About the energy consumption for the formation of fluorine atoms during the dissociation of fluoride and fluoride by an electron beam, Kvant. Elektronika.. - 1983 .-- V. 10 .- P. 428.

52. Khomich V.Yu., Yamshchikov V.A. Fundamentals of creating systems of electric-discharge excitation of high-power CO_2, N_2 and F_2 lasers. - Moscow, Fizmatlit. 2015, 168 p.

53. Burtsev V.A., Vodovozov V.M., Dashuk P.N., Kulakov S.L., Prokopenko V.F., Fomin V.M., Chelnokov L.L. On the simultaneous formation of volumetric and moving discharges of nanosecond duration as applied to pumping gas lasers, Abstracts of VII All-Union Conf. in Physics of low-temperature plasma. Tartu - 1984. - P. 414–416.

54. Apollonov V.V. High-Energy Molecular Lasers. - Springer Series in Optical Sciences, 2016.

55. Mesyats A.G., Osipov V.V., Tarasenko V.F. Pulsed gas lasers. - Moscow, Nauka, 1991 .-- 272 p.

56. Vinogradov A.S., Zimkina T.M. Photoabsorption cross sections of sulfur hexafluoride in the field of soft and ultra-soft x-ray radiation, Optika Spektroskopiya. - 1972. –V. 32. - P. 33-37.

57. Genkin S.A., Karlov N.V., Klimenko K.A. The use of soft x-ray radiation to initiate

a volume discharge in large interelectrode gaps, Pis'ma Zh. Teor. Fiz. –1984. - V. 10. - No. 11. - P. 641-644.

58.   Gaiman V.G., Genkin S.A., Klimenko K.A. Features of the formation of an independent discharge in large interelectrode gaps, Journal of Technical Physics. - 1985. - V. 55. - No. 12. - P. 2347-2353.

59.   Spichkin G.L. Pulsed high-voltage volume discharge in SF6 gas, Zh. Tekh. Fiz. - 1986. - V. 56. - P. 1923-1932.

60.   Brink D.J., Hasson V. Compact megawatt helium-free TEA HF / DF lasers, Journal of Physics E: Scientific Instruments. - 1980. - V. 13. - P. 553-556.

61.   Apollonov V.V., Baytsur G.G., Prokhorov A.M., Firsov K.N. The formation of a volume discharge for pumping CO2 lasers, Kvant. Elektronika.. - 1987. - V. 14. - P. 1940-1942.

62.   Belkov E.P., Dashuk P.N., Kozlov Yu.I., Piskunov A.K., Romanenko Yu.V., Spichkin G.L. The formation of a volume discharge in electronegative gas mixtures, Zh. Tekh. Fiz. - 1982. - V. 52. - P. 1794-1801.

63.   Malyuta D.D., Tolstov V.F. Chemical pulsed HF laser based on a mixture of technical C_3H_8 with SF_6, initiated by an electric discharge, Kvant. Elektronika.. - 1983 - V. 10 .- P. 441-443.

64.   Apollonov V.V., Baytsur G.G., Prokhorov A.M., Semenov S.K., Firsov K.N. The effect of easily ionizable substances on the stability of a volume discharge in working mixtures of CO_2 lasers, Kvant. Elektronika.. - 1988. - V. 15. - P. 553-560.

65.   Wensel R.G. and Arnold G.P. A Double-Discharge-Initiated HF laser, IEEE J. Quantum Electronics. - 1972. - V. 8. - P. 26-27.

66.   Arnold G.P. and Wensel R.G. Improved Performance of an Electrically Initiated HF laser, IEEE J. Quantum Electronics. - 1973. - V. 9. - P. 491-493.

67.   Apollonov V.V., Baytsur G.G., Prokhorov A.M., Firsov K.N. The formation of a self-sustained volume discharge in dense gases at large interelectrode distances, Pisma Zh. Teor. Fiz. - 1985. - V. 11. - No. 20. - P. 1262–1266.

68.   68. Apollonov V.V., Firsov K.N., Kazantsev S.Y., Oreshkin V.F. High-power SSD-based pulse nonchain HF (DF) laser, Proceedings of SPIE, High-Power Laser Ablation. - 1998. - V. 3343. - P. 783-788.

69.   Apollonov V.V., Belevtsev A.A., Kazantsev S.Yu., Sayfulin A.V., Firsov K.N. Self-initiated volume discharge in nonchain HF lasers based on mixtures of SF6 with hydrocarbons, Kvant. Elektronika.. - 2000. - T. 30. - No. 3. - S. 207-214.

70.   Velikanov S.D., Zapolsky A.F., Frolov Yu.N. Physical aspects of the operation of HF and DF lasers with a closed cycle of changing the working medium, Kvant. Elektronika.. - 1997. - T. 24. - No. 1. - S. 11-14.

71.   Harris M.R., Morris A.V., Gorton E.K. Closed-cycle 1-kHz-pulse-repetition-frequency HF (DF) laser, Proceedings of SPIE. - 1998. - V. 3268. - P. 247-251.

72.   Zhou S., Ma L., Huang K. et.al. Experimental investigation on factors influencing output energy stability of non-chain HF laser, High Power Laser and Particle Beams. - 2015. - V. 27. - No. 9. - P. 091001.

73.   Serafetinidest A.A. and Rickwood K.R. Improved performance of small and compact TEA pulsed HF lasers employing semiconductor preionisers, Journal of Physics E: Scientific Instruments. - 1989. - V. 22. - P. 103-107.

74.   Hatch C.B. A compact, resistive electrode HF laser suitable for optical studies of semiconductors, Journal of Physics E: Scientific Instruments. - 1980. - V. 13. - P. 589-591.

75.   Velikanov S.D., Evdokimov P.A., Zapolsky A.F., Kovalev E.V., Pegoev I.N. Features of the formation of a volume discharge in an HF (DF) laser using blade electrodes, Kvant. Elektronika.. - 1998. - V. 25. - No. 10. - P. 925–926.

76.   Andramanov A.V., Kabaev S.A., Lazhintsev B.V., Nor-Arevyan V.A., Pisetskaya A.V., Selemir V.D. High-frequency HF laser with plate electrodes, Kvant. Elektron-

ika.. - 2006. - T. 36. - No. 3. - S. 235–238.

77. 77. Aksenov Yu. N., Borisov V.P., Burtsev Val. V., Velikanov S. D., Voronov S. L., et al. Pulse-periodic DF laser with a generation power of 400 W, Kvant. Elektronika..- 2001. - V. 31. - No. 4. - P. 290-292.

78. 78. Lazarenko V.I. Infrared sources of the mid-IR range, Tenth All-Russian School for Students, Graduate Students, Young Scientists and Specialists in Laser Physics and Laser Technologies, Collection of reports / Ed. S.G. Garanin. - Sarov: FSUE RFNC-VNIIEF. - 2017 .-- S. 133-140.

79. 79. Apollonov V.V., Kazantsev S.Yu., Oreshkin V.F., Firsov K.N. Effective non-chain HF (DF) lasers with high output characteristics, Pis'ma Zh. Teor. Fiz. - 1996. - T. 22. - No. 24. - S. 60-63.

80. 80. Bulaev V.D., Kulikov V.V., Petin V.N., Yugov V.I. An experimental study of a non-chain HF laser using heavy hydrocarbons, Kvant. Elektronika.. - 2001. - T. 31. - No. 3. - S. 218–220.

81. 81. Andreev Yu.M., Velikanov S.D., Elutin A.S., Zapolsky A.F., Konkin D.V., Mikshin S.N. Smirnov S.V., Frolov Yu.N., Schurov V.V. SHG radiation from a DF laser in ZnGeP2, Kvant. Elektronika.. - 1992. - T. 19. - No. 11. - S. 1110.

82. 82. Weis T.A., Goldberg L.S. Singly resonant CdSe parametric oscillator pumped by an HF laser, Applied Physics Letters. - 1974. - V. 24. - P. 389.

83. 83. Garnov S.V., Scherbakov I.A. Laser methods for generating mega-volt terahertz pulses., Usp. Fiz. Nauk. - 2011. - V. 181. - No. 1. - P. 97-102.

84. 84. John H. Smith and Dean W. Robinson Chemical pumping of pure rotational HF lasers, The Journal of Chemical Physics. - 1981. - V. 74. - P. 5111.

85. 85. Geraldine L. Richmonda and George C. Pimentel HF rotational laser emission from the ClF / H2 reaction: Time evolution of the gain, The Journal of Chemical Physics. - 1984. - V. 80. - P. 1162.

86. 86. Molevich N.E., Pichugin S.Yu. Pulsed H2-F2 laser with simultaneous generation of radiation at rotational and vibrational-rotational transitions, Kvant. Elektronika.. - 2011. - V. 41. - No. 5. - P. 427-429.

87. 87. Kozlovsky V.I., Korostelin Yu.V., Landman A.I., Mislavsky V.V., Podmarkov Yu.P., Skasyrsky Y.K., Frolov M.P. Pulsed Fe^{2+}: ZnS laser with smooth wavelength tuning in the region of 3.49-4.65 μm, Kvant. Elektronika.. - 2011.- V. 41. P. 1-3.

88. 88. Kernal J., Fedorov V.V., Gallian A., Mirov S. B., Badikov V.V. 3.9-4.8 mkm gain-switched lasing of Fe: ZnSe at room temperature, Optics Express. - 2005. - V. 13. - No. 26. - P. 10608-10612.

89. Kozlovsky V.I., Akimov V.A., Frolov M.P., Korostelin Yu.V., Landman A.I., Martoitsky V.P., Mislavskii V.V., Podmar'kov Yu.P., Sksyrsky Ya.K., Voronov A.A. "Room-temperature tunable mid-infared lasers on transition-metal doped II-VI compound crystals grown from vapor phase", Physica Status Solidi B. - 2010. - V. 247. - No. 6. - P. 1553-1556.

90. 90. Doroshenko M.E., Jelinkova H., Sulc J., Jelinek M., Nemec M., Basiev T.T., Zagoruiko Y.A., Kovalenko N.O., Gerasimenko A.S., Puzikov V.M. Laser properties of Fe: Cr: Zn1-xMgxSe crystal for tunable mid - infrared laser sources, Laser Physics Letters. - 2012. - V. 9. - No. 4. - P. 301-305.

91. 91. Mirov S.B., Fedorov V.V., Martyshkin D.V., Moskalev I.S., Mirov M.S. and Vasilyev S.V. Progress in Mid-IR Lasers Based on Cr and Fe-Doped II – VI Chalcogenides, IEEE Journal of selected topics in quantum electronics. - 2015. - V. 21. - P. 1601719.

92. 92. Apollonov V.V., Kazantsev S.Yu., Oreshkin V.F., Firsov K.N. Possibilities for increasing the output energy of a nonchain HF (DF) laser, Kvant. Elektronika.. -

1997. - V. 24. - No. 3. - P. 213-215.

93. Apollonov V.V., Kazantsev S.Yu., Oreshkin V.F., Firsov K.N. Non-chain electric discharge HF (DF) laser with high radiation energy, Kvant. Elektronika.. - 1998. - V. 25. - No. 2. - P. 123-125.

94. Apollonov V.V., Kazantsev S.Yu., Oreshkin V.F., Firsov K.N. Independent volume discharge in mixtures of SF_6 with hydrocarbons (carbon deuterides) at large interelectrode distances, II Intern. conf. in plasma physics and plasma technology. (Minsk, ed. NAS of Belarus. - 1997. - S. 150).

95. Apollonov V.V., Kazantsev S.Yu., Oreshkin V.F., Firsov K.N. The stability of the self-sustained volume discharge in mixtures of SF_6 with hydrocarbons (carbon deuterides), VIII Conf. in gas discharge physics (Ryazan, Ryazan publishing house. Radioengineering Acad. - 1996. - P. 7).

96. Apollonov V.V., Kazantsev S.Yu., Oreshkin V.F., Firsov K.N. Features of the development of a self-sustained volume discharge in SF_6 and mixtures of SF_6 with hydrocarbons (carbon deuterides), IX Conf. in gas discharge physics (Ryazan, Ryazan publishing house. Radioengineering Acad. - 1998. - P. 58.).

97. Apollonov V.V., Belevtsev A.A., Firsov K.N., Kazantsev S.Yu., Saifulin A.V. Self-sustained volume discharge in mixtures of SF6 with gydrocarbons, Proceedings of III Int. Conf. Plasma Physics And Plasma Technology, PPPT-3, Minsk, Belarus, 18-22 September 2000. - 2000. - V. 1. - P. 27-30.

98. Apollonov V.V., Belevtsev A.A., Kazantsev S.Yu., Sayfulin A.V., Firsov K.N. Ion-ion recombination in SF_6 and SF_6-C_2H_6 mixtures at high E / N, Kvant. Elektronika. - 2001. - V. 31. - No. 7. - P. 629-633.

99. Apollonov V.V., Kazantsev S.Yu., Sayfulin A.V., Firsov K.N. Characteristics of a discharge in a non-chain HF (DF) laser, Kvant. Elektronika. - 2000. - V. 30. - No. 6. - P. 483-485.

100. Apollonov V.V., Belevtsev A.A., Kazantsev S.Yu., Sayfulin A.V., Firsov K.N. Features of the development of a self-initiating volume discharge in nonchain HF lasers, Kvant. Elektronika. - 2002. - V. 32. - No. 2. - P. 95-100.

101. Belevtsev A.A., Kazantsev S.Yu., Sayfulin A.V., Firsov K.N. Once again on the role of UV illumination in nonchain electric-discharge HF (DF) lasers, Kvant. Elektronika. - 2004. - V. 34. - No. 2. - P. 111-114.

102. Belevtsev A.A., Firsov K.N., Kazantsev S.Yu., Kononov I.G., Shchurov V.V., Velikanov S.D. Stabilization of a self-sustained volume discharge in SF6 and SF6-based mixtures using bulk resistive SiC cathodes, Proc. V Intern. Conf. on Plasma Physics and Plasma Technology, PPPT-5, Minsk, Belarus, September 18-22, 2006 .-- 2006. - V. 1. - P. 58-61.

103. Bychkov Yu, Gortchakov S., Lacour B., Pascuiers S., Puech V., Yastrmsky A.G. Two-step ionization in non-equilibrium SF6 discharges at high current density, Journal of Physics D: Applied Physics. - 2003. - V. 36. - P. 380–384.

104. Belevtsev A.A., Kazantsev S.Yu., Sayfulin A.V., Firsov K.N. Self-initiated volume discharge in iodides for the production of iodine atoms in an oxygen iodine laser, Kvant. Elektronika. - 2003. - V. 33. - No. 6. - P. 489-492.

105. Month G.A. Actons, Usp. Fiz. Nauk. - 1995. - V. 165. - No. 6. - P. 601-626.

106. Korolev Yu.D., Mesyats G.A. Field emission and explosive processes in a gas discharge. - Moscow, Nauka, 1982.- 252 p.

107. Belevtsev A.A., Kazantsev S.Yu., Kononov I.G., Firsov K.N. Characteristics of a self-sustained volume discharge in SF_6-based mixtures / Abstracts at the XXXIV International (Zvenigorod) Conference on Plasma Physics and Controlled Thermonuclear Fusion, Zvenigorod, February 12-16, 2007. - 2007. - P. 175.

108. Apollonov V.V., Firsov K.N., Kazantsev S.Yu., Oreshkin V.F., Saifulin A.V. Scaling up of non-chain HF (DF) laser initiated by self-sustained volume discharge, Proceedings of SPIE. - 2000. - V. 3886.P. 370-381.

109. Bychkov Yu. I. and Yastremskii A. G. Kinetic Processes in the Electric Discharge in SF6, Laser Physics. - 2006. - V. 16. - No. 1. - P. 146–154.

110. Bychkov Yu., Gortchakov S., Yampolskaya S., Yastremskii A. Kinetic Processes in the Electric Discharge in SF6, Proc. SPIE. - 2002. - V. 4047. - P. 106.

111. Yu. I. Bychkov, S. A. Yampolskaya, A. G. Yastremsky. Kinetic processes in an inhomogeneous discharge plasma in SF6-based gas mixtures, Izv. VUZ. Fizika. - 2007. - V. 50. - No. 9. - P. 236-240.

112. Demyanov A.V., Kochetov I.V., Napartovich A.P., Kapitelli M., Longo S. Influence of vibrational kinetics of HCl on the development of micro-instabilities and characteristics of an electric-discharge XeCl laser under conditions of inhomogeneous preionization, Kvant. Elektronika.. - 1995. - V. 22. - No. 7. - P. 673–682.

113. Hilmert H., Schmidt W.F. Electron detachment from negative ions of sulphur hexafluoride-swarm experiments, Journal of Physics D: Applied Physics. - 1991. - V. 24. - P. 915-921.

114. Bychkov Yu., Gortchakov S., Yastremskii A. Homogeneity and stability of volume electrical discharges in gas mixtures based on SF6, Quantum Electronics. - 2000. - V. 30. - P. 733–737.

115. Boychenko A.M., Tkachenko A.N., Yakovlenko S.I. Townsend coefficient and runaway of electrons in an electronegative gas, Pis'ma Zh. Eksper. Teor. Fiz. - 2003. - V. 78. - P. 1223-1227.

116. Christophorou L.G., Olthoff J.K. Electron interactions with SF6, Journal of Physical and Chemical Reference Data. - 2000. - V. 29. - P. 267-330.

117. Radzig A.A., Smirnov B.M. Handbook of atomic and molecular physics. - Moscow, Atomizdat, 1980 .-- 240 p.

118. Massey H.S.W. Negative Ions. Oxford, Cambridge University Press, 1976.

119. Apollonov V.V., Belevtsev A.A., Firsov K.N., Kazantsev S.Yu., Saifulin A.V. Current Density Restriction in Diffuse Channel of a Self-Initiated Volume Discharge in SF6 and SF6-C2H6 mixtures, Proceedings of XXV International Conference on Phenomena in Ionized Gases, ICPIG-2001, Nagoya, Japan, July 17-22, 2001. - V. 1. - P. 247-249.

120. Belevtsev A.A., Firsov K.N., Kazantsev S.Yu, Kononov I.G. Podlesnykh S.V. On the current density limiting effect in SF6-based mixtures, J. Phys. D: Appl. Phys. - 2011. - V. 44. - No. 50. - P. 505502.

121. Belevtsev A.A., Kazantsev S.Yu., Konov I.G., Lebedev A.A., Podlesnykh S.V., Firsov K.N. On the stability of a self-sustained volume discharge in working mixtures of a non-chain electrochemical HF laser, Kvant. Elektronika. - 2011. - V. 41. - No. 8. - P. 703-708.

122. Slovetsky D.I. The mechanisms of chemical reactions in a nonequilibrium plasma - Moscow, Nauka, 1980. - 312 p.

123. Repin P.B. Study of high-power laser systems with nanosecond electrodischarge initiation / Abstract dis. Cand. Phys.-Math. sciences. - M., 1990.

124. Repyev A.G., Repin P.B. Space-time parameters of x-ray radiation of a diffuse atmospheric discharge, Zh. Tekh. Fiz. - 2008. - V. 78. - No. 1. - P. 78-85.

125. Babich LB, Loyko T.V., Zuckerman V.A. High-voltage nanosecond discharge in dense gases at high overvoltages, developing in the runaway mode of electrons, Usp. Fiz. Nauk. - 1990. - V. 160. - No. 7. - P. 49-82.

126. Illenberger E., Smirnov B.M. Electron attachment to free and bound molecules,

Usp. Fiz. Nauk. - 1998. - T=V. 168. - P. 731-766.

127. Belevtsev A.A., Firsov K.N., Kazantsev S.Yu., Kononov I.G. The Burning Voltage of a Self-Sustained Volume Discharge in CO_2-laser irradiated SF_6 and Mixtures of SF_6 with C_2H_6, Proc. SPIE. - 2006. - V. 6053. - P. 109-118.

128. Belevtsev A.A., Firsov K.N., Kazantsev S.Yu., Kononov I.G. A self-sustained volume discharge in vibrationally excited strongly electronegative gases, Journal of Physics D: Applied Physics. - 2004. - V. 37.P. 1759-1764.

129. Belevtsev A.A., Firsov K.N., Kazantsev S.Yu., Kononov I.G. The influence of vibrational excitation on the burning voltage of a pulsed electric discharge in HF laser working media, Appl. Phys.B. - 2006. - V. 82. - No. 3. - P. 455-462.

130. Makarov G.N. Selective processes of IR excitation and dissociation of molecules in gasdynamically cooled jets and flows, Usp. Fiz. Nauk. - 2005. - V. 175. - S. 41-84.

131. Molin Yu.N., Panfilov VN, Petrov A.K. Infrared photochemistry. - N .: Nauka, 1985 .-- 253 p.

132. Belevtsev A.A., Kazantsev S.Yu., Kononov I.G., Firsov K.N. Obtaining the temperature dependence of the critical electric field in SF6 and mixtures of SF6 with C2H6 by laser gas heating, Kvant. Elektronika. - 2007. - V. 37. - No. 10. - S. 985-988.

133. Belevtsev A.A., Firsov K.N., Kazantsev S.Yu., Kononov I.G. On the temperature dependence of the critical reduced electric field in SF_6 and mixtures of SF_6 with C_2H_6, J. Phys. D: Appl. Phys. - 2007. - V. 40. - No. 5. - P. 1368–1375.

134. Belevtsev A.A., Firsov K.N., Kazantsev S.Yu., Kononov I.G. Dynamics of a self-sustained volume discharge in laser-shock-disturbed SF6-based mixtures, J. Phys. D: Appl. Phys. - 2009. - V. 42. - No. 11. - P. 115202.

135. Belevtsev A.A., Kazantsev S.Yu., Kononov I.G., Firsov K.N. Self-contained volume discharge in gas mixtures based on SF6 during the development of shock-wave perturbations of a medium initiated by a pulsed CO2 laser, Kvant. Elektronika. - 2006. - V. 36. - No. 7. - P. 646-652.

136. Datskos P.G., Christophorou L.G., Carter J.G. Temperature dependence of electron attachment and detachment in SF_6 and c-C4F$_6$, The Journal of Chemical Physics. - 1993. - V. 99. - P. 8607-8616.

137. Eliasson B., Schade E. In Proc. XIII Intern. Conf. on Phenomena in Ionized Gases, Leipzig, Germany. - 1977 .-- P. 409.

138. Shade E. In Proc. XVII Intern. Conf. on Phenomena in Ionized Gases, Invited Lectures, Budapesht, Hungary. - 1985 .-- P. 277.

139. Alexandrov D.A., Alexandrov N.A., Bazelyan E.M., Konchakov A.M. Channel evolution of a long leader in air with a sharp change in the discharge current, Fiz. plazmy. - 2003. - V. 29. - P. 182.

140. Belevtsev A.A., Firsov K.N., Kazantsev S.Yu., Kononov I.G. The critical reduced electric fields in SF6 at high gas temperatures, J. Phys. D: Appl. Phys. - 2008. - V. 41. - No. 4. - P. 045201.

141. Chervy B., Gleizes A., Razafinimanana M. Thermodynamic properties and transport coefficients in SF6-Cu mixtures at temperatures of 300-30000 K and pressures of 0.1-1 MPa, Journal of Physics D: Applied Physics. - 1994. - V. 27. - P. 1193.

142. Babichev A.P., Babushkina N.A., Bratkovsky A.M. In: Physical quantities: reference book - Moscow, Energoizdat, 1991. - 1232p.

143. Urquijo J. Swarm studies on elementary processes and ion – molecule reactions in low temperature plasmas, Plasma Sources Science and Technology. - 2002. - V. 11. - P. 3A.

144. Karlov N.V., Kirichenko N.A., Lukyanchuk B.S. Laser thermochemistry - Moscow, Nauka, 1992. - 296 p.

145. Christophorou L.G. Olthoff J.K. Advances In Atomic, Molecular, and Optical Physics. - 2001. - V. 44. - P. 59-98.
146. Haas R.A. Plasma Stability of Electric Discharges in Molecular Gases, Physical Review A. - 1973. - V. 8. - P. 1017.
147. Nighan W.L., Wiegand W.J. Influence of negative-ion processes on steady-state properties and striations in molecular gas discharges, Physical review A. - 1974. - V. 10. - P. 922.
148. Nighan W.L., In: Principles of laser plasmas, Chapter 7. Ed. G. Bekefi, N.Y., 1976.
149. Napartovich A.P., Starostin A.N. Mechanisms of instability of a glow discharge of high pressure. In: Plasma Chemistry, Issue 6, Moscow: Atomizdat. - 1979.- C. 153-208.
150. Coutts J. The halogen donor depletion instability-current pulse-shape effects, Journal of Physics D: Applied Physics. - 1988. - V. 21. - C. 255-259.
151. Osborne M.R., Hutchinson M.R. Long pulse operation and premature termination of a high power discharge pumped XeCl laser, Journal of Applied Physics. - 1986. - V. 59. - P. 711.
152. Osborne M.R. Rare-gas-halide discharge stability, Applied Physics B. - 1988 .- V. 45. P. 285.
153. Garscaden A., Kushner M.J. Plasma physics issues in gas discharge laser development, IEEE Transactions on Plasma Science. - 1991. - V. 19. - P. 1015-1031.
154. Belevtsev A.A., Firsov K.N., Kazantsev S.Yu., Kononov I.G. Electron detachment instability and self-organization in strongly electronegative polyatomic gases, J. Phys. D: Appl. Phys. - 2009. - V. 42. - No. 21. - P. 215205.
155. Yeletsky A.V.in: Plasma Chemistry Vol. 9 (Under the editorship of B.M. Smirnov) - Moscow, Energoatomizdat, 1982. - 151 p.
156. Velikhov E.P, Kovalev A.S., Rakhimov A.T. Physical phenomena in a gas discharge plasma. - M.: Science, Fizmatlit, 1987.
157. Belevtsev A.A., Biberman L.M., News of the Academy of Sciences of the USSR, Energetika i transport. - 1976. - V. 3. - P. 74.
158. Volkov A.F., Kogan Sh.M. Physical phenomena in semiconductors with negative differential conductivity, Usp. Fiz. Nauk. - 1968.- V. 96.- S. 633-672.
159. Shell E. Self-organization in semiconductors. - Moscow: Mir, 1991 . -459 p.
160. Apollonov V.V., Kazantsev S.Yu., Oreshkin V.F., Sayfulin A.V., Firsov K.N. Independent volume discharge for initiating wide-aperture nonchain HF (DF) lasers, Izv. RAN. Ser. Fiz. - 2000. - V. 64. - No. 7. - P. 1439-1443.
161. Apollonov V.V., Belevtsev A.A., Firsov K.N., Kazantsev S.Yu., Saifulin A.V. High-Power Pulse And Pulse-Periodic Non-chain HF (DF) Lasers, Proceedings of SPIE. - 2002. - V. 4747. - P. 31-43.
162. Ignatiev A.B., Kazantsev S.Yu., Kononov I.G., Marchenko V.M., Feofilaktov V.A., Firsov K.N. On the possibility of controlling the wavefront of a wide-aperture HF (DF) laser using Talbot interferometry, Kvant. Elektronika. - 2008. - V. 38. - No. 1. - S. 69-72.
163. Bulaev V.D., Gusev V.S., Kazantsev S.Yu., Kononov I.G., Lysenko S.L., Morozov Yu.B., Poznyshev A.N., Firsov K.N. High-energy pulse discharge periodic HF laser, Kvant. Elektronika. - 2010. - V. 40. - No. 7. - P. 615-618.
164. Velikanov S.D., Garanin S.G., Domazhirov A.P., Efanov E.M., Efanov M.V., Kazantsev S.Yu., Kodola B.E., Komarov Yu.N., Kononov I.G., Podlesnykh S.V., Sivachev A.A., Firsov K.N., Schurov V.V., Yarin P.M. Powerful electric-discharge HF laser with a solid-state pump generator, Kvant. Elektronika. - 2010. - V. 40. - No. 5. - P. 393-396.

165. Velikanov S.D., Domazhirov A.P., Zaretsky N.A., Kazantsev S.Yu., Kononov I.G., Kromin A.A., Podlesnykh S.V., Sivachev A.A., Firsov K.N., Kharitonov S.V., Tsykin V.S., Schurov V.V., Yutkin I.M. Powerful repetitively pulsed HF (DF) laser with a solid-state pump generator, Kvant. Elektronika. - 2015. - V. 45. - No. 11. - S. 989-992.

166. Lacur B., Brunet H., Besaucelle H. and Garnol C. High average power XeCl excimer laser, Proc. SPIE. - 1992. - V. 1810. - P. 498.

167. Brunet H., Lacur B., Legentil M., Mizzi S., Pasquiers S. and Puech V. Theoretical and experimental studies of phototriggered discharges in argon and neon, Journal of Applied Physics. - 1990. - V. 68. - P. 4474.

168. Bespalov V.I., Pasmanik G.A. Nonlinear optics and adaptive laser systems - Moscow, Nauka, 1986. - 133 p.

169. Abrosimov Yu.M., Drozhbin Yu.A., Morozov Yu.B. Prokopenko V.E., Semenov A.K., Semenov V.B., Favorov L.N. Measurement of the divergence of pulsed laser radiation by the focal spot method using a mirror wedge, Izmerit. Tekhnika.- 1982. - V. 11. - P. 30-32.

170. Vasiliev L.A. Shadow methods - Moscow: Nauka, 1968 .-- 400 p.

171. Optical production control. Ed. Malakary D., - Moscow, Mashinostronie,1985. - 400 p.

172. Koryakovsky A.S., Marchenko V.M., Prokhorov A.M. The diffraction theory of the Talbot-interferometry method and the diagnostics of wide-aperture wave fronts, Tr. IOFAN. - 1987. - V. 7. - P. 33-92.

173. Bulaev V.D., Gusev V.S., Lysenko S.L., Morosov Yu.B., Poznyshev A.N., Firsov K.N., Kazantsev S.Yu., Kononov I.G., High power pulse-periodical electrochemical HF laser, Chinese Optics. - 2011. - V. 4. - No. 1. - P. 26-30.

174. Vartapetov S.K., Gryaznov O.V., Malashin M.V., Moshkunov S.I., Nebogatkin S.V., Khasaya R.R., Khomich V.Yu., Yamshchikov V.A. An electric-discharge VUV laser with a solid-state pump generator, Kvant. Elektronika. - 2009. - V. 39. - P. 714–718.

175. Borisov V., Khristoforov O., Kirykhin Yu., Vinokhodov A., Demin A., Eltzov A., Proc. SPIE. - 2001. - V. 4184. - P. 348.

176. 176. Tarasenko V.F., Baksht E.H., Kunts S.E., Panchenko A.N. Pumping of discharg gas lasers by generators with inductive energy storage and semiconducting opening switch, Proc. SPIE. - 1999. - V. 3612. - P. 22-31.

177. 177. Panchenko A.N., Orlovsky V.M., Tarasenko V.F., Baksht E.Kh. Efficient generation modes of an HF laser pumped by a non-chain chemical reaction initiated by an independent discharge, Kvant. Elektronika. - 2003. - V. 33. - No. 5. - P. 401–407.

178. 178. FID GmbH. http://www.fidtech.com

179. 179. Butsykin I.L., Velikanov S.D., Evdokimov P.A. et al. Pulse-periodic DF laser with a pulse repetition rate of up to 1200 Hz and an average power of ~25 W, Kvant. Elektronika. - 2001. - V. 31. - P. 957–961.

180. 180. Velikanov S.D., Elutin A.S., Kudryashov E.A., Pegoev I.N., Sinkov S.N., Frolov Yu.N. The use of a DF laser for the analysis of hydrocarbons in the atmosphere, Kvant. Elektronika. - 1997. - V. 24. - P. 279–282.

181. Zuev V.V., Zuev V.E. Laser environmental monitoring of gas components of the atmosphere - Moscow, Nauka, 1992. - 182 p.

182. Handbook of infrared technology. Ed. Wolf W., Cisis G. - Moscow, Mir. 1995 .- 606 p.

183. Tonniben A., Wanner J., Rothe K.W., Walther H. Application of a CW chemical laser for remote pollution monitoring and process control, Applied Physics. - 1979. - V. 18. - No. 3. - P. 297-304.

184. Trautman M. Determination of gas concentrations and temperature in the exhaust of

a power plant with a cw chemical HF laser, Applied Physics B. - 1986. - V. 40. - No. 1. - P. 29-33.

185. Mezheris R. Laser remote sensing. Mosocw. Mir, 1987 .-- 550 p.
186. Zakharov V.M. Lidars and climate research. Leningrad, Gidrometeoizdat, 1990 .-- 320 p.
187. Kazantsev S.Yu., Kononov I.G., Kossy I.A., Tarasova N.M., Firsov K.N. Ignition of a combustible gas mixture in a closed volume initiated by a freely localized laser spark, Plasma Physics. - 2009. - T. 35. - No. 3, - S. 281-288.
188. Artem'ev K.V., Berezhetskaya N.K., Kazantsev S.Yu., Kononov I.G., Kossyi I.A., Popov N.A., Tarasova N.M., Filimonova E.A., Firsov K.N. Fast combustion waves and chemi-ionization processes in a flame initiated by a powerful local plasma source in a closed reactor, Philosophical Transactions A. - 2015. - V. 373. - No. 2048. - P. 20140334.
189. Old man A.M. Titova N.S. On the possibility of intensifying chain reactions in combustible mixtures upon excitation of electronic states of O_2 molecules by laser radiation, Dokl. Akad. Nauk. - 2003. - V. 191. - No. 4. P. 471-477.
190. Bravyi B.G., Makarov E.F., Chernyshev Yu.A. Optical pumping of mixtures of N_2O-He and N_2O-CO_2-He by radiation from a pulsed multi-frequency HF laser to create active media for amplification of powerful ultrashort pulses of a ten-micron range, Kvant Elektronika. - 2005. - V. 35. P. 987–992.
191. Alexandrov B.S., Arsenjev A.V., Azarov M.A., Burtsev A.P., Drozdov V.A., Mashendzhinov V.I., Revich V.E., Sudarikov V.V., Tonkov M.V., Troshchinenko G.A. Increase of efficiency of optical pumping of a broadband CO2 laser amplifier as a result of the use of a multicomponent active medium, Proc. SPIE. - 2003. - V. 5120. - P. 551-556.
192. Azarov M.A., Alexandrov B.S., Arsenjev A.V., Burtsev A.P., Drozdov V.A., Kalinovsky V.V., Mashendzhinov V.I., Mishchenko G.M., Troshchinenko G.A. ($12CO_2$ + $13CO_2$ + $14N_2O$ + $15N_{14}NO$) -active media of high pressure with optical pumping by HF laser radiation, XVI International Symposium on Gas Flow, Chemical Lasers, and High-Power Lasers. - International Society for Optics and Photonics, 2007.- T. 6346. - S. 63462B.
193. Gao M.H., et al. Chinese Optics. - 2013. - V. 6. - No. 6. - P. 810-817.
194. Andreev S.N., Firsov K.N., Kazantsev S.Yu., Kononov I.G. and Samokhin A.A. Explosive Boiling of Water Induced by the Pulsed HF-Laser Radiation, Laser Physics. - 2007. - V. 17. - No. 6. - P. 834-841.
195. Andreev S.N., Il'ichev N.N., Firsov K.N., Kazantsev S.Yu., Kononov I.G., Kulevskii L.A. and Pashinin P. P. Generation of an Electrical Signal upon the Interaction of Laser Radiation with Water Surface, Laser Physics. - 2007. - V. 17. - No. 8. - P. 1041-1052.
196. Andreev S.N., Kazantsev S.Yu., Kononov I.G., Pashinin P.P., Firsov K.N. Generation of an electrical signal by the interaction of HF laser radiation with the bottom surface of a water column, Kvant Elektronika. - 2010. - V. 40. - No. 8. - P. 716-719.
197. Zotov S.D., Kazantsev S.Yu., Kudryavtsev E.M., Kuznetsov A.A., Lebedev A.A., Firsov K.N. Measurement of the refractive index in a wave excited in water by a laser pulse, PTE. - 2017. - No. 6. - P. 83–89.
198. Andreev S.N., Kazantsev S.Yu., Kononov I.G., Pashinin P.P., Firsov K.N. Temporal structure of an electric signal during the interaction of HF laser radiation with the bottom surface of a water column, Kvant Elektronika. - 2009. - V. 39. - No. 8. -P. P. 179-184.
199. Galkin A.A., Lunin V.V. Water in sub and supercritical states - a universal medium for the implementation of chemical reactions, Usp. Khimii. - 2005. - V. 74. - No.

1. - P. 24-40.

200. Dolgayev. S.I., Karasev M.E., Kulevsky L.A., Simakin A.V., Shafeev G.A., Dissolution in a supercritical fluid as a mechanism of laser ablation of sapphire, Kvant Elektronika. - 2001. - V. 31. - No. 7. - P. 593–596.

201. Skribanowitz N., Hermann I.P., MacGillivray M.S., Feld M.S. Observation of Dicke Superradiance in Optically Pumped HF Gas, Physical review letters. - 1973. - V. 30. - P. 309.

202. Andreev A.V., Emelyanov V.I., Ilyinsky Yu.A. Collective spontaneous emission, Usp. Fiz. Nauk. - 1980. - V. 131. - No. 4. - P. 653-694.

203. Gavrishchuk E.M., Kazantsev S.Yu., Kononov I.G., Rodin S.A., Firsov K.N. High-energy ZnSe: Fe^{2+} laser operating at room temperature, Kvant Elektronika. - 2014. - V. 44. - No. 6. - P. 505-506.

204. Gavrishchuk E.M., Ikonnikov V.B., Kazantsev S.Yu., Kononov I.G., Rodin S.A., Savin D.V., Timofeeva N.A., Firsov K.N. Scaling the energy characteristics of a ZnSe: Fe^{2+} polycrystal laser at room temperature, Kvant Elektronika. - 2015. - V. 45. - No. 9. - P. 823-827.

205. Firsov K.N., Gavrishchuk E.M., Kazantsev S.Yu., Kononov I.G., Maneshkin A.A., Mishchenko G.M., Nefedov S.M., Rodin S.A., Velikanov S.D., Yutkin I.M., Zaretsky N.A., Zotov E.A. Spectral and temporal characteristics of a ZnSe: Fe^{2+} laser pumped by a non-chain HF (DF) laser at room temperature, Laser Phys. Lett. - 2014. - V. 11. - No. 12. - P. 125004.

206. Firsov K.N., Gavrishchuk E.M., Ikonnikov V.B., Kazantsev S.Yu., Kononov I.G., Kotereva T.V., Savin D.V., Timofeeva N.A. Energy and spectral characteristics of a room-temperature pulsed laser on a ZnS: Fe^{2+} polycrystal, Laser Phys. Lett. - 2016. - V. 13. - No. 4. - P. 145004.

207. Firsov K.N., Gavrishchuk E.M., Ikonnikov V.B., Kazantsev S.Yu., Kononov I.G., Kotereva T.V., Savin D.V. and Timofeeva N.A. Room-temperature laser on a ZnSe:Fe^{2+} polycrystal with undoped faces, excited by an electrodischarge HF laser, Laser Phys. Lett. - 2016. - V. 13. - No. 5. - P. 055002.

208. Firsov K.N., Gavrishchuk E.M., Ikonnikov V.B., Kazantsev S.Yu., Kononov I.G., Kotereva T.V., Savin D.V. and Timofeeva N.A. Room-temperature laser on a ZnS: Fe^{2+} polycrystal with a pulse radiation energy of 0.6 J, Laser Phys. Lett. - 2016. - V. 13. - No. 6. - P. 065003.

209. Dormidonov A.E., Firsov K.N., Gavrishchuk E.M., Ikonnikov V.B., Kazantsev S.Yu., Kononov I.G., Kotereva T.V., Savin D.V., Timofeeva N.A. High-efficiency room-temperature ZnSe: Fe2 + laser with a high pulsed radiation energy, Applied Physics B. - 2016. - V. 122. - P.211.

210. Velikanov S.D., Zaretsky N.A., Zotov E.A., Kazantsev S.Yu., Kononov I.G., Korostelin Yu.V., Maneshkin A.A., Firsov K.N., Frolov M.P., Yutkin I.M. ZnSe: Fe^{2+} laser with a radiation energy of 1.2 J at room temperature, Kvant Elektronika. - 2016. - V. 46. - No. 1. - P. 11-12.

211. Velikanov S.D., Dormidonov A.E., Zaretsky N.A., Kazantsev S.Yu., Kozlovsky V.I., Kononov I.G., Korostelin Yu.V., Maneshkin A.A., Podmarkov Yu.P., Skasyrsky Y.K., Firsov K.N., Frolov M.P., Yutkin I.M. "ZnS: Fe^{2+} single crystal laser excited at room temperature by an electric-discharge HF laser", Kvant Elektronika. - 2016. - V. 46. - No. 9. - P. 769-771.

212. Gavrishuk E., Ikonnikov V., Kotereva T., Savin D., Rodin S., Mozhevitina E., Avetisov R., Zykova M., Avetissov I., Firsov K., Kazantsev S., Kononov I., Yunin P. Growth of high optical quality zinc chalcogenides single crystals doped by Fe and Cr by the solid phase recrystallization technique at barothermal treatment, Journal

of Crystal Growth. - 2017. - V. 468. - P. 655–661.

213. Velikanov S.D., Zaretsky N.A., Zakhryapa A.V., Ikonnikov V.B., Kazantsev S.Yu., Kononov I.G., Maneshkin A.A., Mashkovsky D.A., Saltykov E.V., Firsov K.N., Chuvatkin R.S., Yutkin I.M. Pulse-periodic Fe:ZnSe laser with an average radiation power of 20 W at room temperature of a polycrystalline active element, Kvant. Elektronika. - 2017. - V. 47. - No. 4. - P. 303–307.

214. Gavrishuk E., Zykova M., Mozhevitina E., Avetisov R., Ikonnikov V., Savin D., Firsov K., Kazantsev S., Kononov I., Avetissov I. Investigations of Nanoscale Defects in Crystalline and Powder ZnSe Doped With Fe for Laser Application, Phys. Status Solidi A. - 2017. - V. 215. - No. 4. - P. 1700457.

215. Firsov K.N., Kazantsev S.Yu., Kononov I.G., Sirotkin A.A., Gavrishchuk E.M., Ikonnikov V.B., Rodin S.A., Savin D.V., Timofeeva N.A. CVD-grown Fe^{2+}:ZnSe polycrystals for laser applications, Laser Phys. Lett. - 2017. - V. 14. - No. 5. - P. 055805.

216. Alekseev E.E., Firsov K.N., Kazantsev S.Yu., Kononov I.G., Rogalin V.E. Nonlinear Absorption of Non-Chain HF Laser Radiation in Germanium, Physics of Wave Phenomena. - 2017. - V. 25. - No. 4. - P. 280–288.

217. Alekseev E.E., Kazantsev S.Yu., Kononov I.G., Rogalin V.E., Firsov K.N. Two-photon absorption of radiation from a nonchain HF laser in germanium single crystals, Optika Spektrospokiya. - 2018. - V. 124. - No. 6. - P. 790–794.

218. Gal A.V., Dodonov A.A., Rusanov V.D., Shiryaevsky V.L., Sholin G.V. The effect of the chemical composition of the hydrogen-containing component RH of the working medium and the initiation method on the parameters of a pulsed chemical SF_6 – RH laser, Quantum Electronics. - 1992. - V. 19.- No. 2. - P. 159–161.

219. Gal' A.V., Rusanov V.D., Shiryaevsky V.L., Sholin G.V. Spectral characteristics of radiation from a pulsed chemical HF laser with electron-beam initiation on an SF_6 – HI – H2 mixture, Kvant. Elektronika. - 1992. - V. 19. - No. 11. - P. 1055-1057.

220. Garnov S.V., Mikhailov V.A., Serov R.V., Smirnov V.A., Tsvetkov V.B., Scherbakov I.A. "Investigation of the possibility of creating a multikilowatt solid-state laser with multichannel diode pumping based on optically dense active media", Kvant Elektronika. - 2007. - V. 37. - No. 10. - P. 910–915.

221. Khanin Ya.I. Basics of laser dynamics. - Moscow, Nauka, 1999 .-368 p.

222. Jonathan W. Evans, Ronald W. Stites, Thomas R. Harris. Increasing the performance of an Fe: ZnSe laser using a hot isostatic press, Optical Materials Express. - 2017. - V. 7. - No. 12. - P. 4296-4303.

223. Sean McDaniel, Adam Lancaster, Ronald Stites, Fiona Thorburn, Ajoy Kar, Gary Cook. Cr: ZnSe guided wave lasers and materials, Proc. SPIE. - 2017. - V. 10082. - P. 100820D.

Index

A

Arkad'ev–Marx generator 44, 163, 180
attachment factor 20

C

C_2H_5OH 57
C_2H_6 7, 8, 19, 21, 23, 24, 26, 34, 37, 38, 39, 45, 46, 48, 57, 59, 60, 62, 63, 64, 74, 75
C_3H_7OH 57
carbon deuterides 93, 163
cathode spot 9, 36, 39, 42, 46, 49, 50, 54, 56, 57, 58, 59, 60, 61, 62
chemical HF(DF) laser 1

D

discharge gap 2, 4, 6, 22, 25, 27, 28, 29, 31, 35, 36, 37, 38, 39, 40, 41, 42, 43, 44, 45, 46, 47, 49, 51, 52, 56, 58, 62, 63, 64, 66, 68, 70, 71, 72, 73, 74, 76, 79, 97, 100, 101, 102, 103, 107, 108, 111, 112, 113, 116, 117, 118, 119, 120, 121, 122, 124, 125, 126, 128, 130, 131, 132, 138, 139, 141, 143, 145, 146, 147, 148, 149, 151, 118, 170, 172, 173, 174, 176, 178, 180, 181

E

effect
  ballo-electric effect 187
  Talbot effect 157, 158
electron drift velocity 22
excitation threshold 74

F

Fitch generator circuit 150

I

ion–ion recombination 69, 93